HOLE'S

HUMAN ANATOMY
&PHYSIOLOGY

DAVID SHIER
WASHTENAW COMMUNITY COLLEGE

JACKIE BUTLER
GRAYSON COUNTY COMMUNITY COLLEGE

RICKI LEWIS
THE UNIVERSITY AT ALBANY

WCB **Wm. C. Brown Publishers**

Dubuque, IA   Bogota   Boston   Buenos Aires   Caracas   Chicago
Guilford, CT   London   Madrid   Mexico City   Sydney   Toronto

 **A Times Mirror Company**

Cover Photo credit: © Mark Lewis/Tony Stone Images

The credits section for this book begins on page 237 and
is considered an extension of the copyright page.

A Times Mirror Company

ISBN 0–697–25381-3

Printed in the United States of America by Wm. C. Brown Communications, Inc.,
2460 Kerper Boulevard, Dubuque, IA 52001

10   9   8   7   6   5   4   3   2   1

This Student Study Art Notebook is a gratis ancillary to assist students in note taking during lectures. On each page, there are one, two, or sometimes three figures faithfully reproduced from the textbook. Each figure also corresponds to one of the 300+ acetates available to instructors who adopt this textbook.

The intention is to place the acetate art in front of students (via the notebook) as the instructor uses the overhead during lectures. The advantage to the student is that he/she will be able to see all labels clearly, and take meaningful notes without having to make hurried sketches of the acetate figure.

The pages of the Art Notebook are perforated and three-hole punched, so they can be removed and placed in a personal binder for specific study and review, or to create space for additional notes.

DIRECTORY OF NOTEBOOK FIGURES

TO ACCOMPANY
SHIER ET AL., HOLE'S HUMAN ANATOMY AND PHYSIOLOGY, 7E.

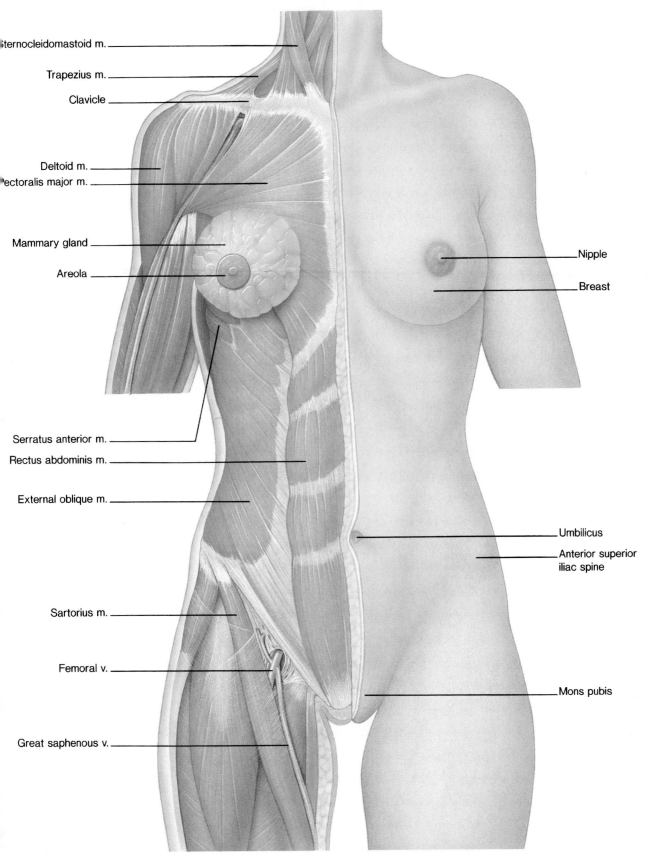

Sternocleidomastoid m.

Trapezius m.

Clavicle

Deltoid m.

Pectoralis major m.

Mammary gland

Areola

Nipple

Breast

Serratus anterior m.

Rectus abdominis m.

External oblique m.

Umbilicus

Anterior superior
iliac spine

Sartorius m.

Femoral v.

Mons pubis

Great saphenous v.

Human Torso, Anterior Surface and Superficial Muscles
Reference Plate 1

1

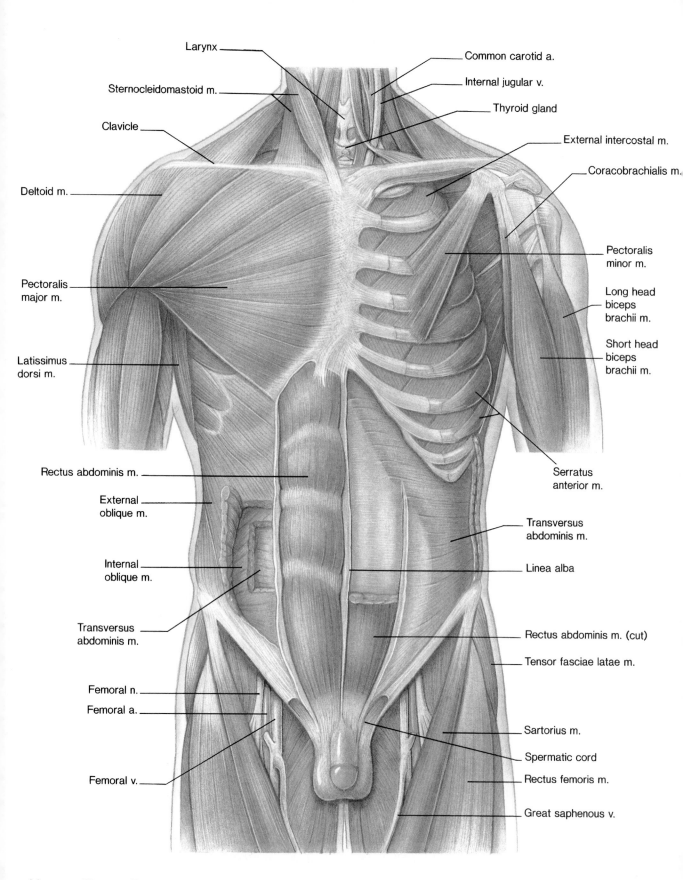

Larynx

Common carotid a.

Internal jugular v.

Sternocleidomastoid m.

Thyroid gland

Clavicle

External intercostal m.

Coracobrachialis m.

Deltoid m.

Pectoralis minor m.

Long head biceps brachii m.

Pectoralis major m.

Short head biceps brachii m.

Latissimus dorsi m.

Serratus anterior m.

Rectus abdominis m.

External oblique m.

Transversus abdominis m.

Internal oblique m.

Linea alba

Transversus abdominis m.

Rectus abdominis m. (cut)

Tensor fasciae latae m.

Femoral n.

Femoral a.

Sartorius m.

Spermatic cord

Femoral v.

Rectus femoris m.

Great saphenous v.

Human Torso, Deeper Muscle Layers
Reference Plate 2

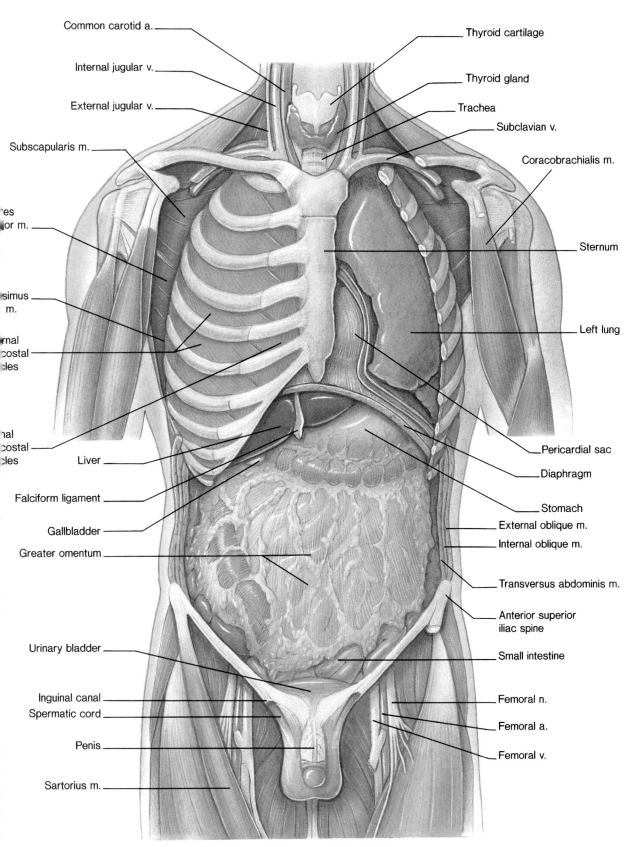

Common carotid a.

Internal jugular v.

External jugular v.

Subscapularis m.

es
jor m.

simus
m.

rnal
costal
les

al
costal
les

Liver

Falciform ligament

Gallbladder

Greater omentum

Urinary bladder

Inguinal canal

Spermatic cord

Penis

Sartorius m.

Thyroid cartilage

Thyroid gland

Trachea

Subclavian v.

Coracobrachialis m.

Sternum

Left lung

Pericardial sac

Diaphragm

Stomach

External oblique m.

Internal oblique m.

Transversus abdominis m.

Anterior superior
iliac spine

Small intestine

Femoral n.

Femoral a.

Femoral v.

Human Torso, Abdominal Viscera
Reference Plate 3

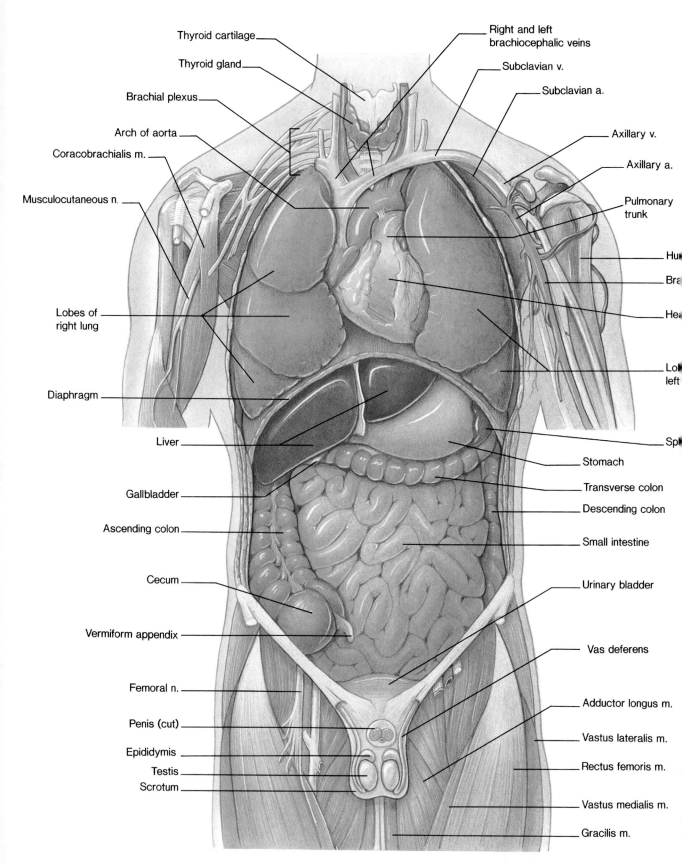

Thyroid cartilage

Thyroid gland

Brachial plexus

Arch of aorta

Coracobrachialis m.

Musculocutaneous n.

Lobes of
right lung

Diaphragm

Liver

Gallbladder

Ascending colon

Cecum

Vermiform appendix

Femoral n.

Penis (cut)

Epididymis

Testis

Scrotum

Right and left
brachiocephalic veins

Subclavian v.

Subclavian a.

Axillary v.

Axillary a.

Pulmonary
trunk

Hu

Bra

Hea

Lo
left

Sp

Stomach

Transverse colon

Descending colon

Small intestine

Urinary bladder

Vas deferens

Adductor longus m.

Vastus lateralis m.

Rectus femoris m.

Vastus medialis m.

Gracilis m.

Human Torso, Thoracic Viscera
Reference Plate 4

4

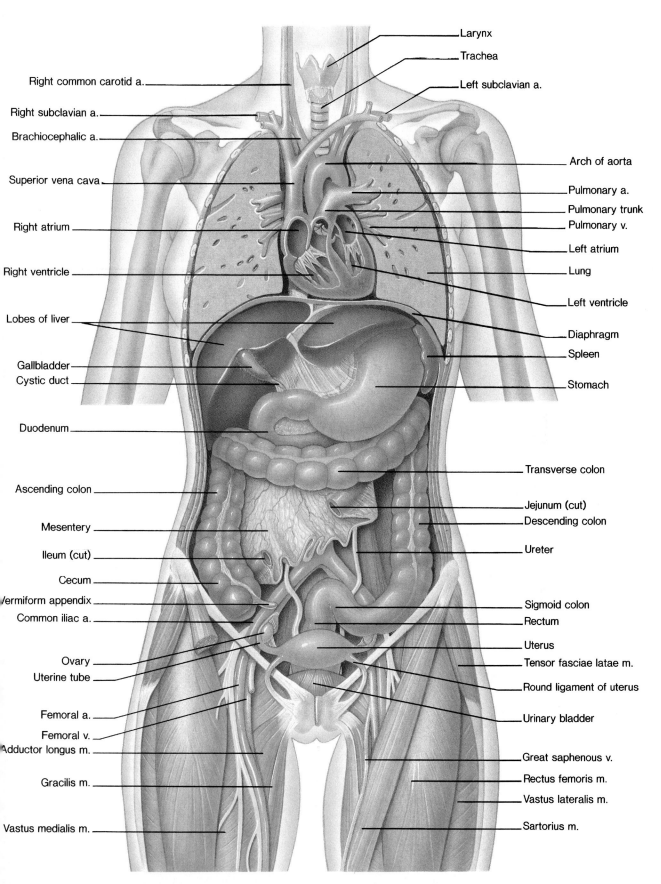

Right common carotid a.

Right subclavian a.

Brachiocephalic a.

Superior vena cava

Right atrium

Right ventricle

Lobes of liver

Gallbladder
Cystic duct

Duodenum

Ascending colon

Mesentery

Ileum (cut)

Cecum

Vermiform appendix
Common iliac a.

Ovary
Uterine tube

Femoral a.
Femoral v.
Adductor longus m.

Gracilis m.

Vastus medialis m.

Larynx

Trachea

Left subclavian a.

Arch of aorta

Pulmonary a.
Pulmonary trunk
Pulmonary v.

Left atrium

Lung

Left ventricle

Diaphragm

Spleen

Stomach

Transverse colon

Jejunum (cut)
Descending colon

Ureter

Sigmoid colon

Rectum

Uterus
Tensor fasciae latae m.

Round ligament of uterus

Urinary bladder

Great saphenous v.

Rectus femoris m.

Vastus lateralis m.

Sartorius m.

Human Torso, Lungs, Heart, and Small Intestine Sectioned
Reference Plate 5

5

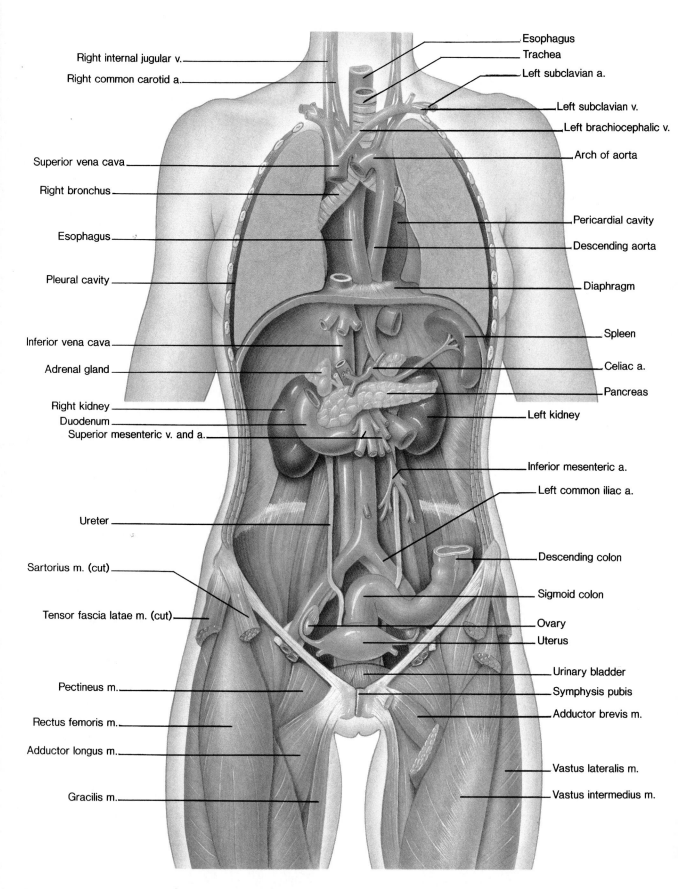

Right internal jugular v.
Right common carotid a.
Superior vena cava
Right bronchus
Esophagus
Pleural cavity
Inferior vena cava
Adrenal gland
Right kidney
Duodenum
Superior mesenteric v. and a.
Ureter
Sartorius m. (cut)
Tensor fascia latae m. (cut)
Pectineus m.
Rectus femoris m.
Adductor longus m.
Gracilis m.

Esophagus
Trachea
Left subclavian a.
Left subclavian v.
Left brachiocephalic v.
Arch of aorta
Pericardial cavity
Descending aorta
Diaphragm
Spleen
Celiac a.
Pancreas
Left kidney
Inferior mesenteric a.
Left common iliac a.
Descending colon
Sigmoid colon
Ovary
Uterus
Urinary bladder
Symphysis pubis
Adductor brevis m.
Vastus lateralis m.
Vastus intermedius m.

Human Torso, Heart, Stomach and Part of Intestine and Lungs Removed
Reference Plate 6

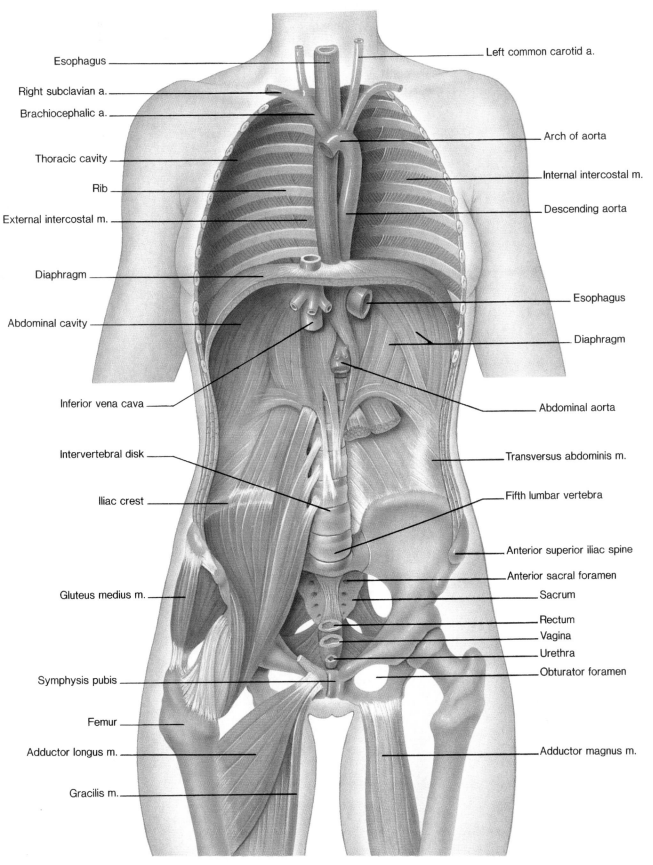

Esophagus

Right subclavian a.

Brachiocephalic a.

Thoracic cavity

Rib

External intercostal m.

Diaphragm

Abdominal cavity

Inferior vena cava

Intervertebral disk

Iliac crest

Gluteus medius m.

Symphysis pubis

Femur

Adductor longus m.

Gracilis m.

Left common carotid a.

Arch of aorta

Internal intercostal m.

Descending aorta

Esophagus

Diaphragm

Abdominal aorta

Transversus abdominis m.

Fifth lumbar vertebra

Anterior superior iliac spine

Anterior sacral foramen

Sacrum

Rectum

Vagina

Urethra

Obturator foramen

Adductor magnus m.

Human Torso, Visceral Organs Removed
Reference Plate 7

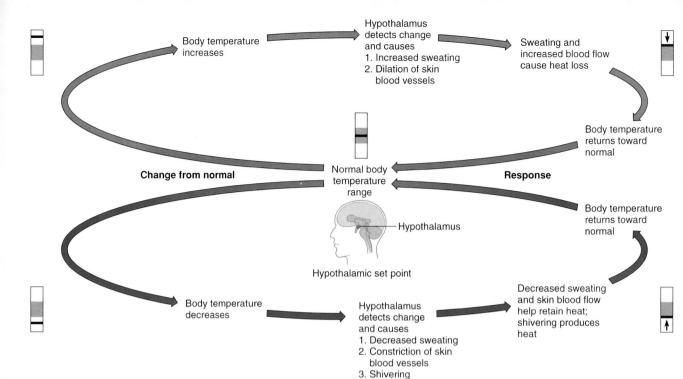

Homeostatic Mechanism
Figure 1.5

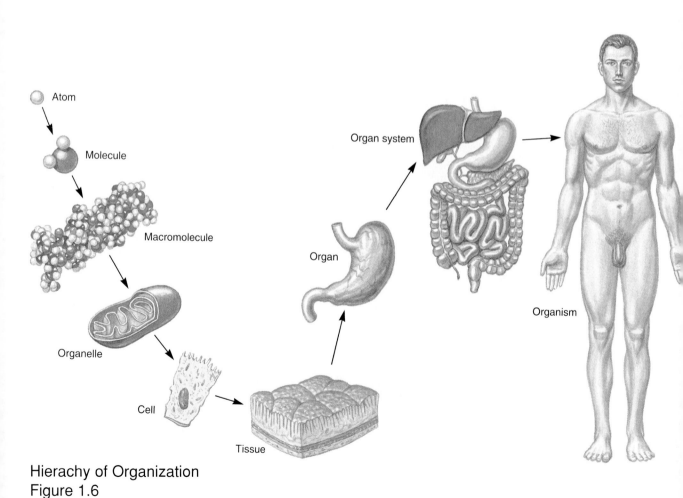

Hierachy of Organization
Figure 1.6

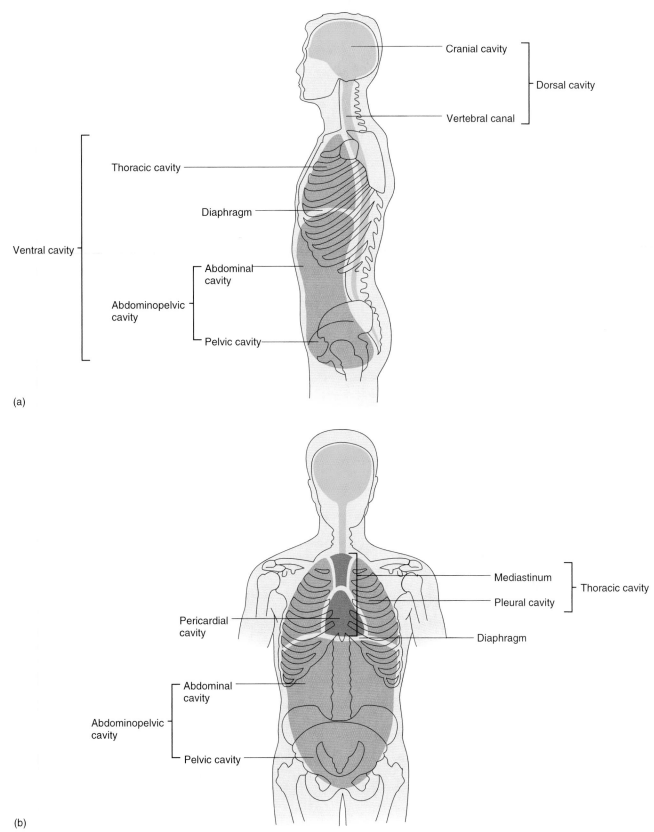

Major Body Cavities
Figure 1.7

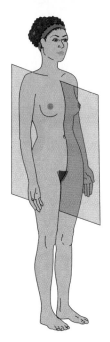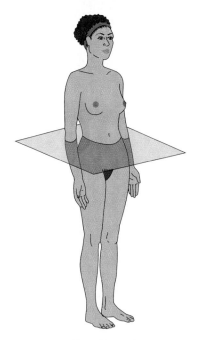

| Sagittal plane
(median plane) | Transverse plane
(horizontal plane) | Coronal plane
(frontal plane) |

Body Planes
Figure 1.17

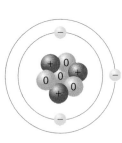

| Hydrogen (H) | Helium (He) | Lithium (Li) |

Organization of an Atom
Figure 2.2

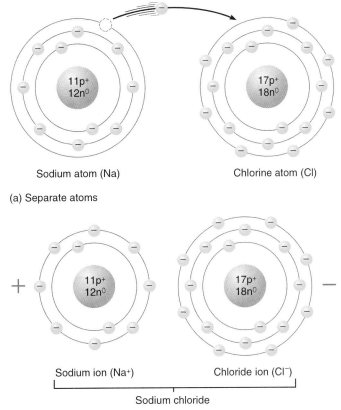

Sodium atom (Na) Chlorine atom (Cl)

(a) Separate atoms

$+$ Sodium ion (Na$^+$) Chloride ion (Cl$^-$) $-$

Sodium chloride

(b) Bonded ions

Ionic Bond
Figure 2.3

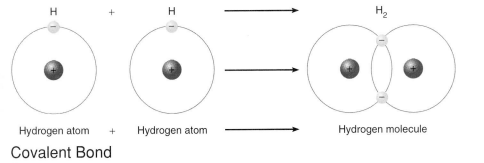

H $+$ H $\longrightarrow$ H$_2$

Hydrogen atom $+$ Hydrogen atom $\longrightarrow$ Hydrogen molecule

Covalent Bond
Figures 2.4

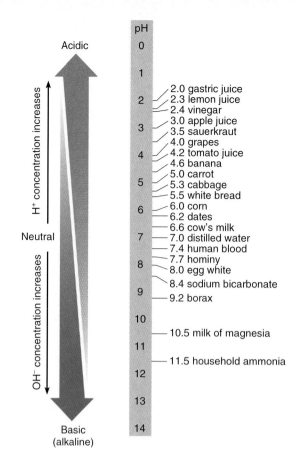

pH of Common Substances
Figure 2.8

The pH scale from 0 (Acidic) to 14 (Basic alkaline), with Neutral at 7.

H⁺ concentration increases (toward Acidic)

OH⁻ concentration increases (toward Basic)

2.0 gastric juice
2.3 lemon juice
2.4 vinegar
3.0 apple juice
3.5 sauerkraut
4.0 grapes
4.2 tomato juice
4.6 banana
5.0 carrot
5.3 cabbage
5.5 white bread
6.0 corn
6.2 dates
6.6 cow's milk
7.0 distilled water
7.4 human blood
7.7 hominy
8.0 egg white
8.4 sodium bicarbonate
9.2 borax
10.5 milk of magnesia
11.5 household ammonia

(a)

(b)

(c)

Structure of Glucose
Figure 2.9

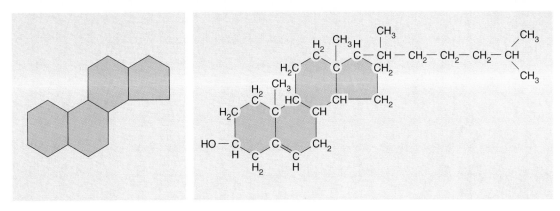

(a) Structure of a steroid   (b) Cholesterol

General Structure of a Steroid
Figure 2.14

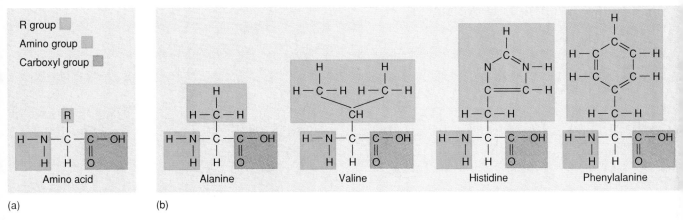

(a)   (b)

Structure of an Amino Acid
Figure 2.15

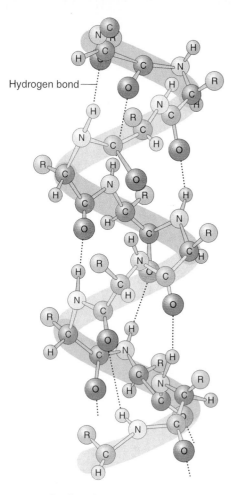

Portion of a protein molecule

Secondary Structure of a Protein
Figure 2.17

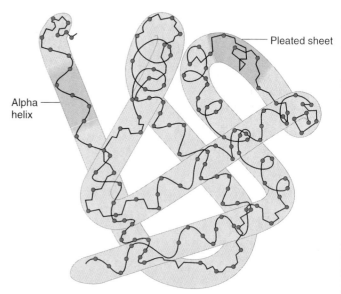

Tertiary Structure of a Protein
Figure 2.18

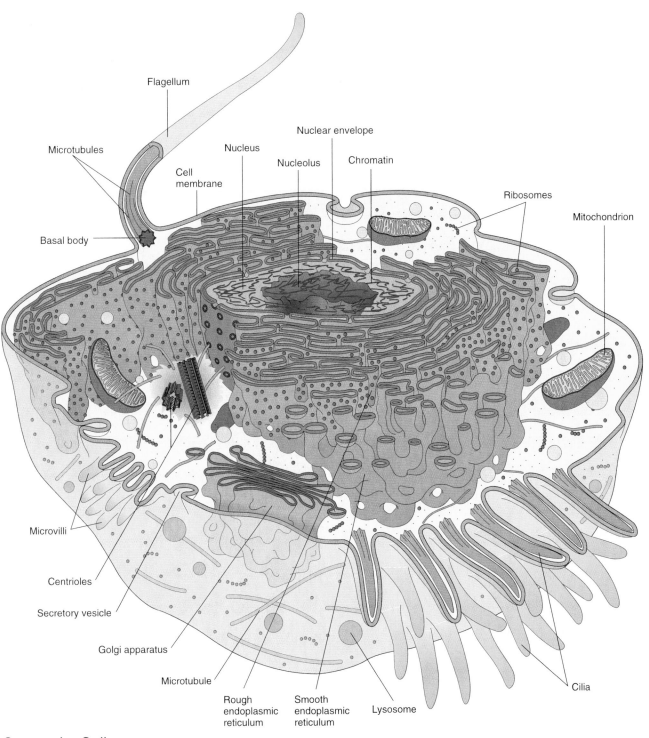

Flagellum

Microtubules

Cell
membrane

Basal body

Nucleus

Nuclear envelope

Nucleolus

Chromatin

Ribosomes

Mitochondrion

Microvilli

Centrioles

Secretory vesicle

Golgi apparatus

Microtubule

Rough
endoplasmic
reticulum

Smooth
endoplasmic
reticulum

Lysosome

Cilia

Composite Cell
Figure 3.3

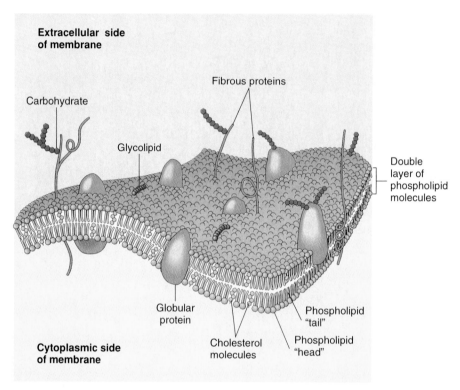

Cell Membrane
Figure 3.7

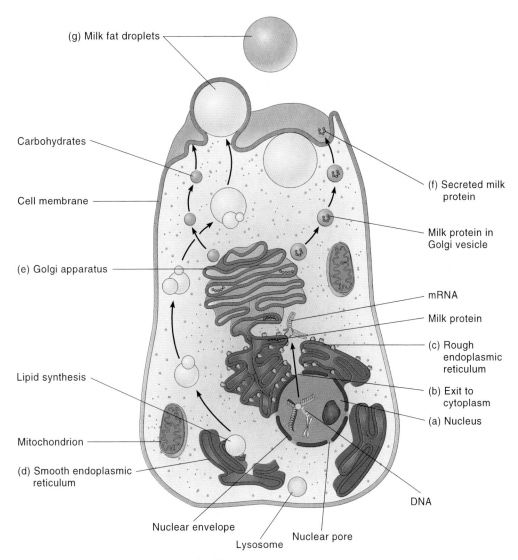

(g) Milk fat droplets

Carbohydrates

Cell membrane

(e) Golgi apparatus

Lipid synthesis

Mitochondrion

(d) Smooth endoplasmic
reticulum

Nuclear envelope

Lysosome

Nuclear pore

(f) Secreted milk
protein

Milk protein in
Golgi vesicle

mRNA

Milk protein

(c) Rough
endoplasmic
reticulum

(b) Exit to
cytoplasm

(a) Nucleus

DNA

Secretory Processes of a Cell
Figure 3.12

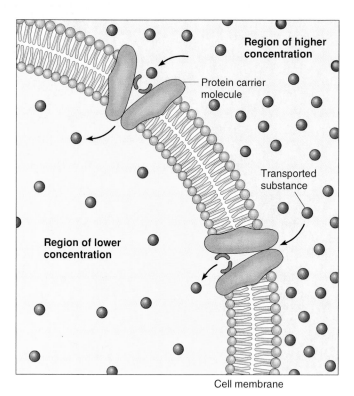

Region of higher
concentration

Protein carrier
molecule

Transported
substance

Region of lower
concentration

Cell membrane

Facilitated Diffusion
Figure 3.23

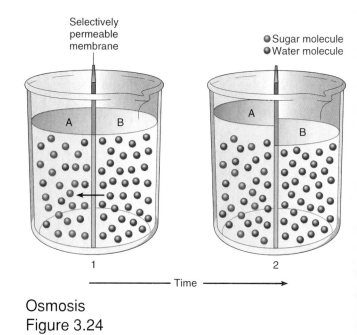

Selectively
permeable
membrane

Sugar molecule
Water molecule

A B

A

B

1

2

Time

Osmosis
Figure 3.24

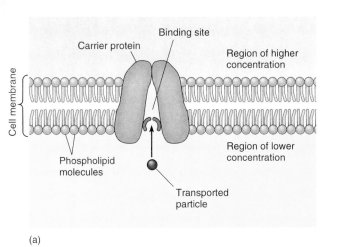
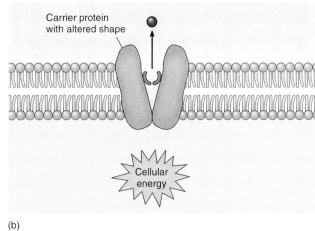

(a)

(b)

Active Transport
Figure 3.28

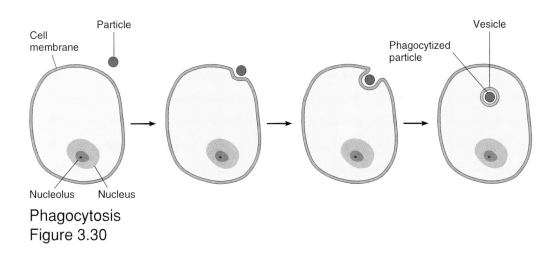

Phagocytosis
Figure 3.30

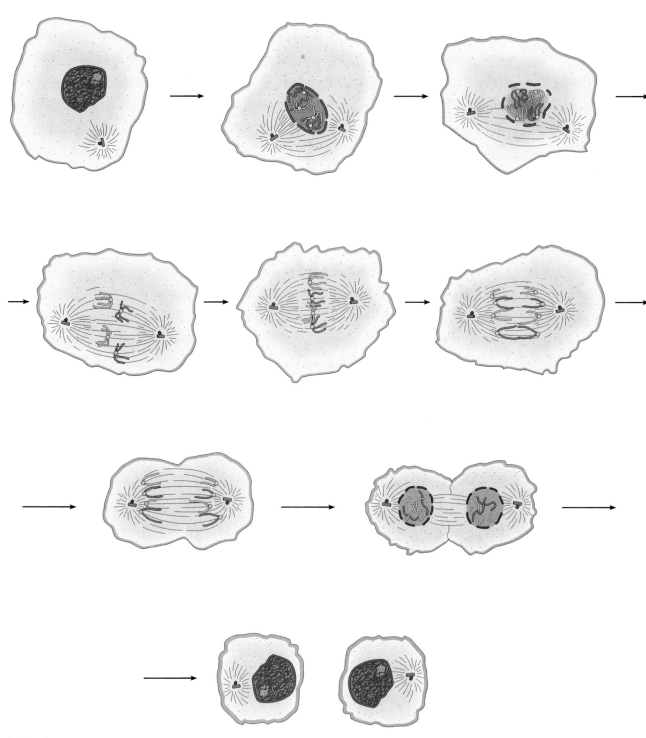

Mitosis
Figure 3.35

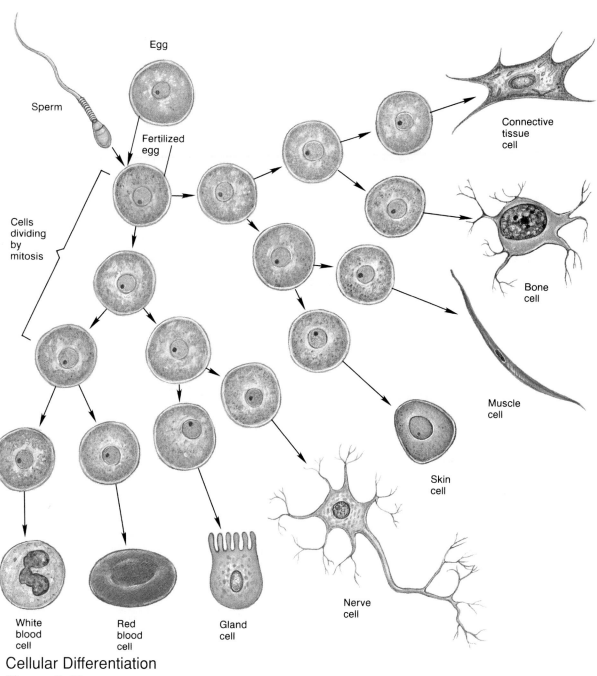

Egg

Sperm

Fertilized egg

Cells dividing by mitosis

Connective tissue cell

Bone cell

Muscle cell

Skin cell

Nerve cell

White blood cell

Red blood cell

Gland cell

Cellular Differentiation
Figure 3.42

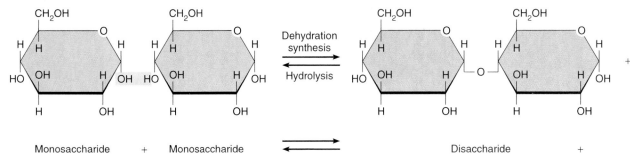

Monosaccharide   +   Monosaccharide   ⇌   Disaccharide   +

Dehydration Synthesis Forms Disaccharide
Figures 4.1

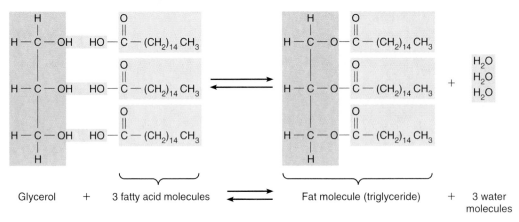

Glycerol   +   3 fatty acid molecules   ⇌   Fat molecule (triglyceride)   +   3 water molecules

Dehydration Synthesis Forms Fat
Figure 4.2

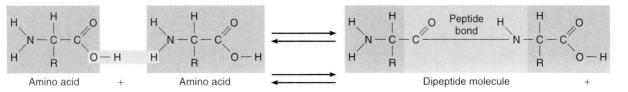

Dehydration Synthesis/Peptide Bond
Figure 4.3

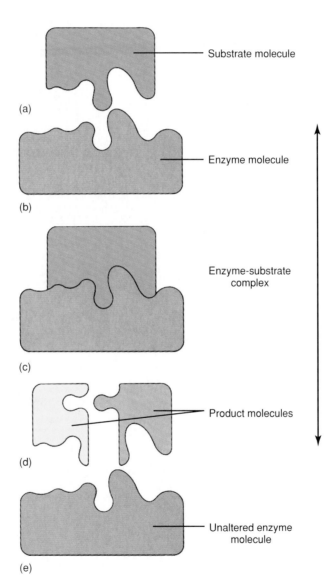

Substrate molecule

Enzyme molecule

(a)

(b)

Enzyme-substrate
complex

(c)

Product molecules

(d)

Unaltered enzyme
molecule

(e)

Enzyme-Substrate Interaction
Figure 4.5

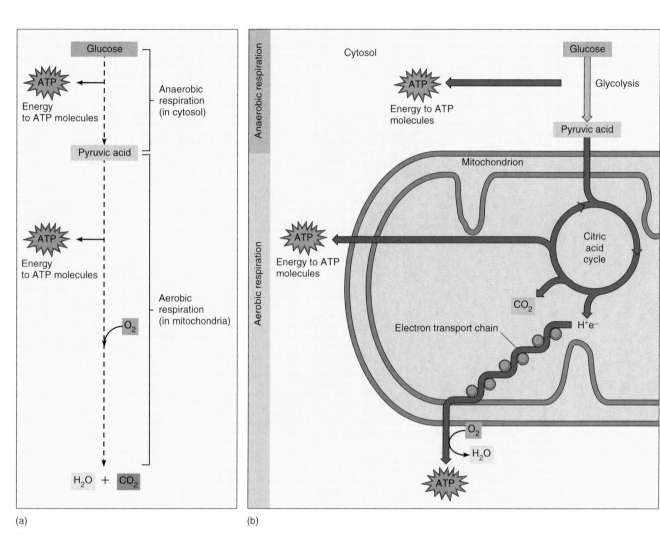

(a)

(b)

Cellular Metabolism
Figure 4.6

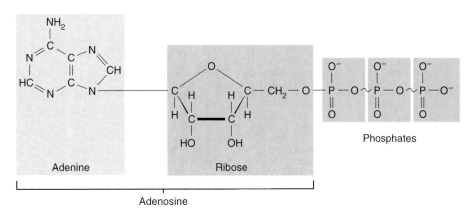

Adenine

Ribose

Phosphates

Adenosine

ATP

Figure 4.7

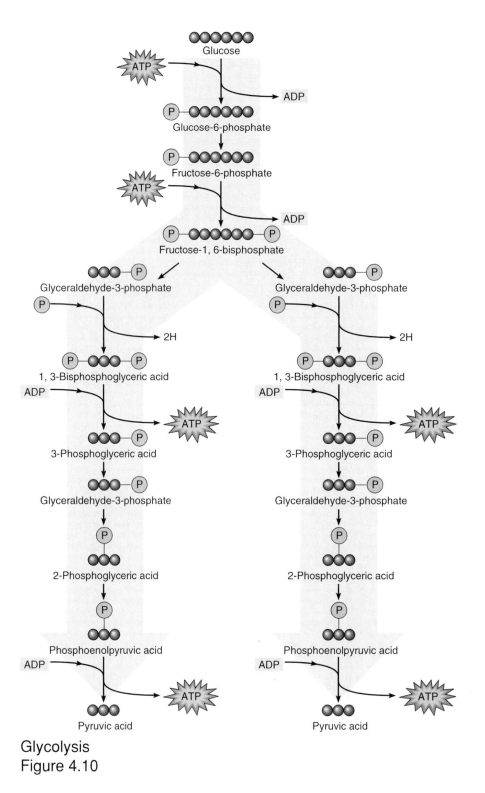

Glycolysis
Figure 4.10

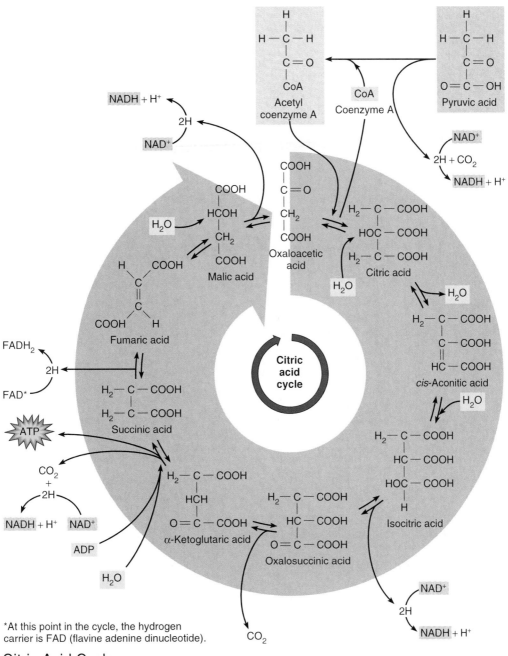

*At this point in the cycle, the hydrogen carrier is FAD (flavine adenine dinucleotide).

Citric Acid Cycle
Figure 4.11

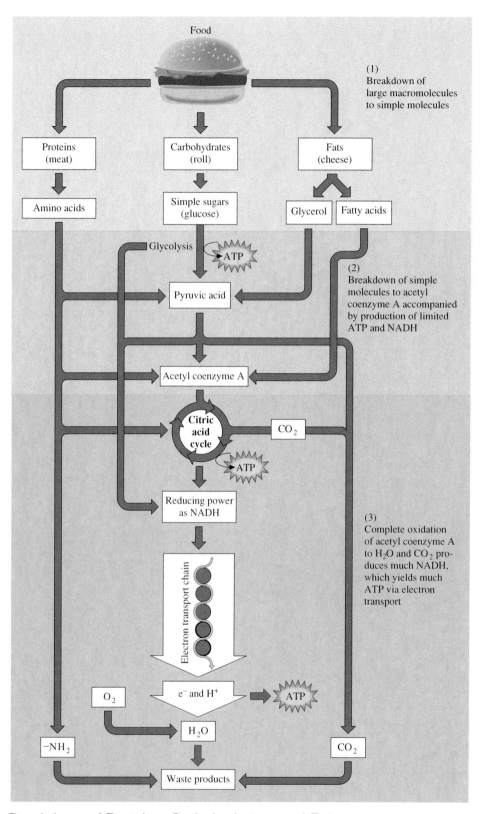

Breakdown of Proteins, Carbohydrates, and Fats
Figure 4.16

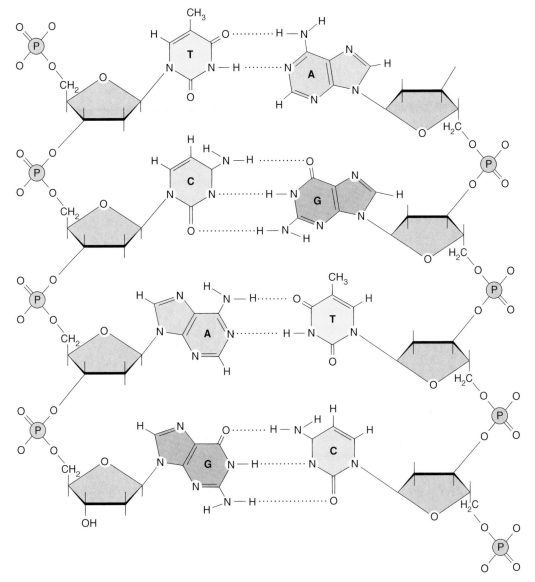

Base-Pairing of DNA Bases
Figure 4.21

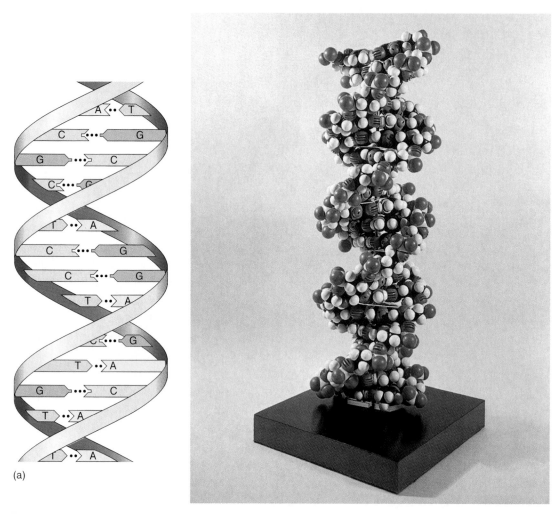

Portion of a DNA Double Helix
Figure 4.22

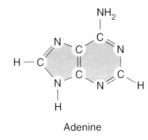

Adenine

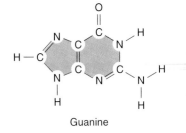

Uracil

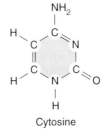

Cytosine

Guanine

RNA Bases
Figure 4.23

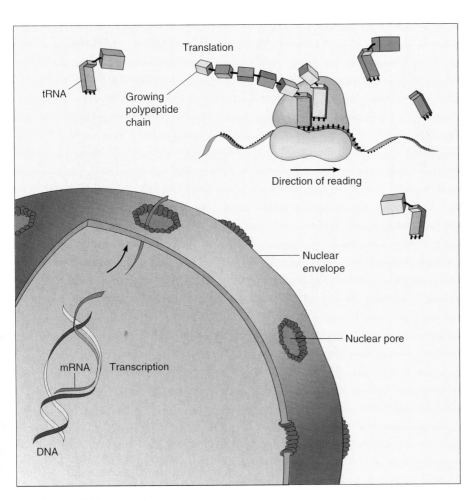

DNA-RNA Transcription
Figure 4.25

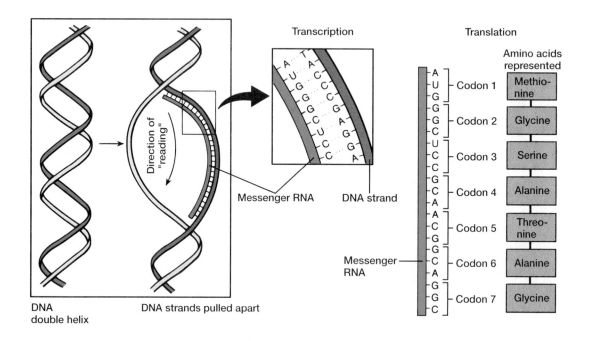

Protein Synthesis
Figure 4.26

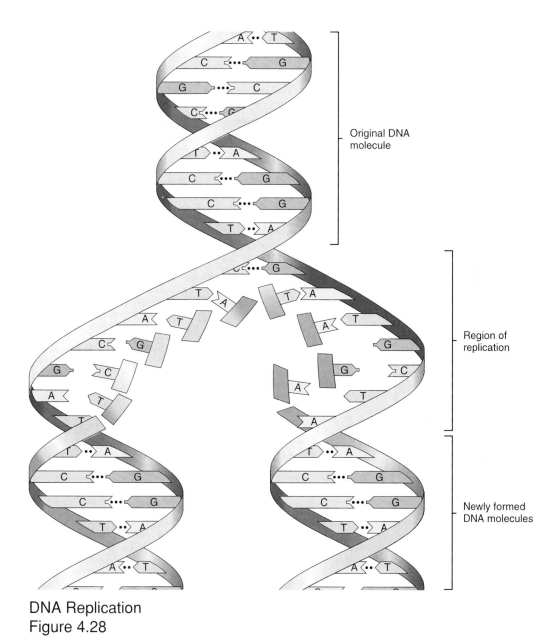

Original DNA
molecule

Region of
replication

Newly formed
DNA molecules

DNA Replication
Figure 4.28

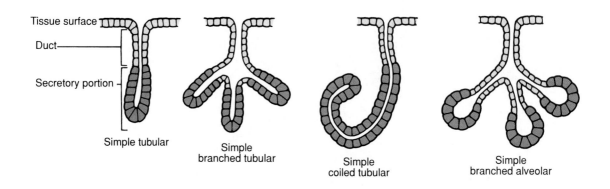

Tissue surface

Duct

Secretory portion

Simple tubular

Simple
branched tubular

Simple
coiled tubular

Simple
branched alveolar

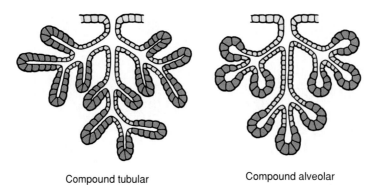

Compound tubular

Compound alveolar

Exocrine Glands
Figure 5.10

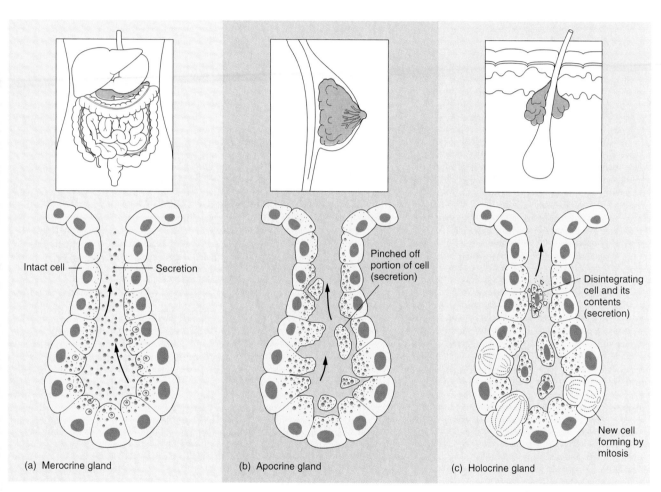

Intact cell — Secretion

(a) Merocrine gland

Pinched off portion of cell (secretion)

(b) Apocrine gland

Disintegrating cell and its contents (secretion)

New cell forming by mitosis

(c) Holocrine gland

Merocrine Glands
Figure 5.11

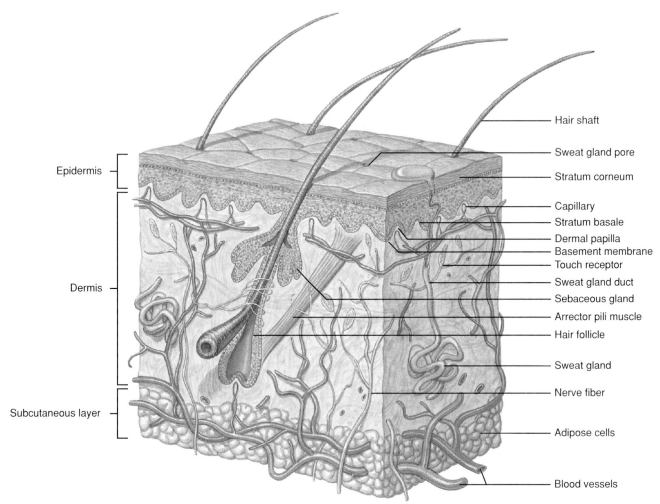

Epidermis

Dermis

Subcutaneous layer

Hair shaft

Sweat gland pore

Stratum corneum

Capillary

Stratum basale

Dermal papilla

Basement membrane

Touch receptor

Sweat gland duct

Sebaceous gland

Arrector pili muscle

Hair follicle

Sweat gland

Nerve fiber

Adipose cells

Blood vessels

Skin Section
Figure 6.2

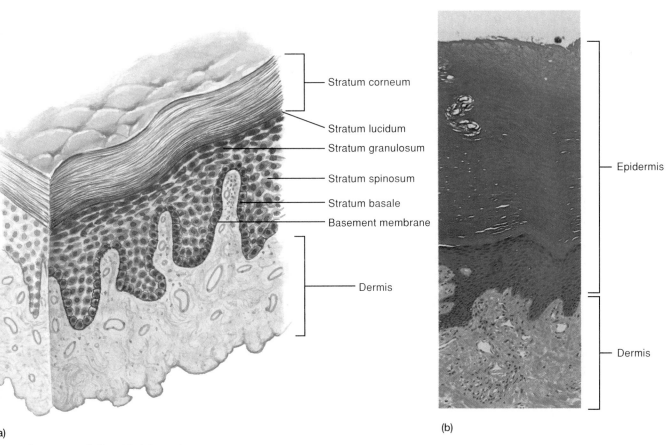

Stratum corneum

Stratum lucidum

Stratum granulosum

Stratum spinosum

Stratum basale

Basement membrane

Dermis

Epidermis

Dermis

(a)

(b)

Layers of the Epidermis
Figure 6.3

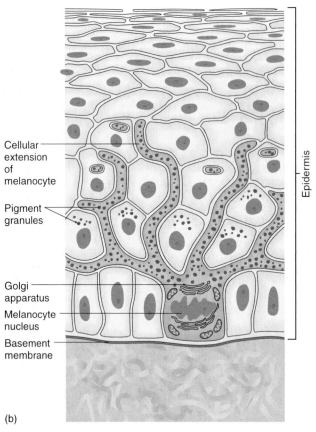

Cellular extension of melanocyte

Pigment granules

Golgi apparatus

Melanocyte nucleus

Basement membrane

Epidermis

(b)

A Melanocyte
Figure 6.5b

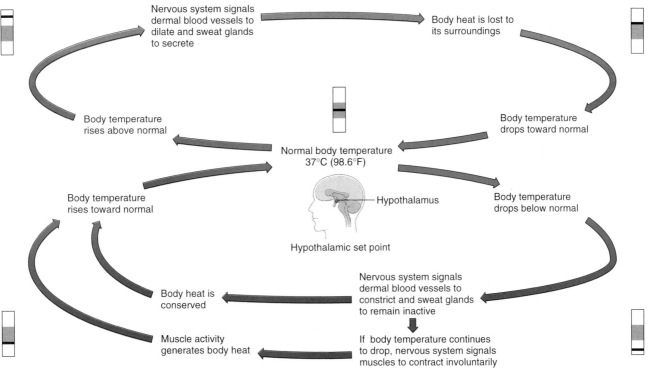

Nervous system signals
dermal blood vessels to
dilate and sweat glands
to secrete

Body heat is lost to
its surroundings

Body temperature
rises above normal

Body temperature
drops toward normal

Normal body temperature
37°C (98.6°F)

Hypothalamus

Body temperature
rises toward normal

Body temperature
drops below normal

Hypothalamic set point

Body heat is
conserved

Nervous system signals
dermal blood vessels to
constrict and sweat glands
to remain inactive

Muscle activity
generates body heat

If body temperature continues
to drop, nervous system signals
muscles to contract involuntarily

Body Temperature Regulation
Figure 6.12

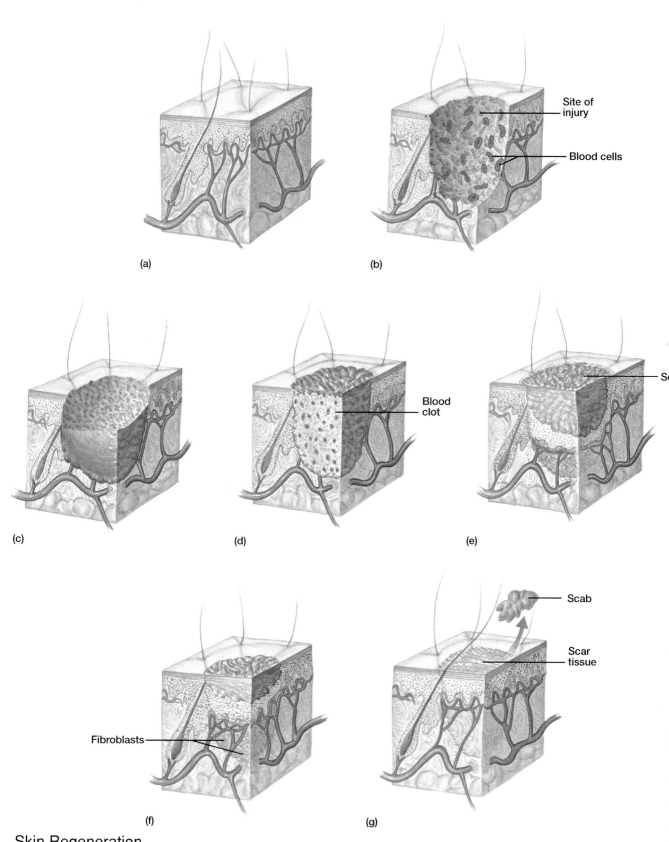

(a)

(b) Site of injury · Blood cells

(c)

(d) Blood clot

(e) S•

(f) Fibroblasts

(g) Scab · Scar tissue

Skin Regeneration
Figure 6.14

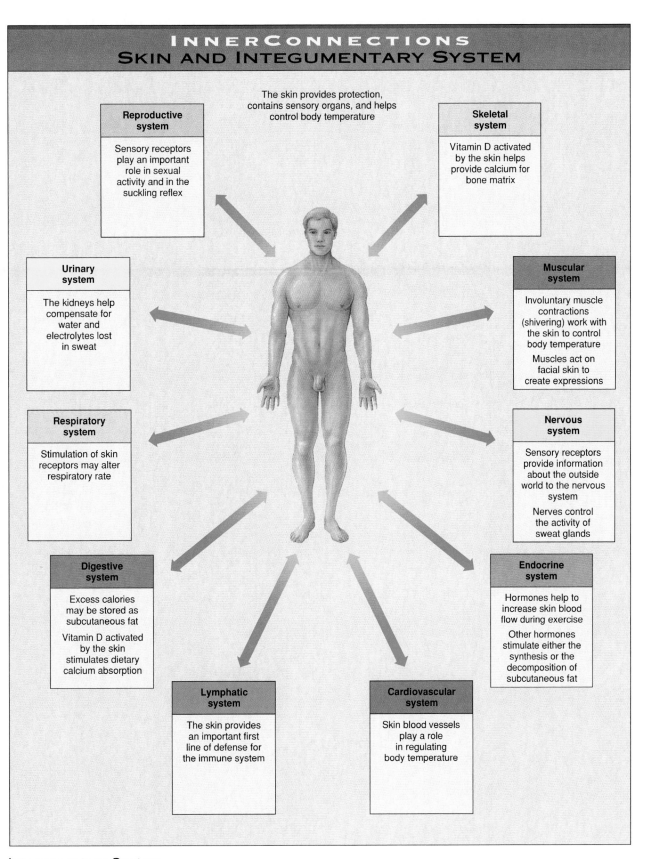

InnerConnections
Skin and Integumentary System

The skin provides protection, contains sensory organs, and helps control body temperature

Reproductive system

Sensory receptors play an important role in sexual activity and in the suckling reflex

Skeletal system

Vitamin D activated by the skin helps provide calcium for bone matrix

Urinary system

The kidneys help compensate for water and electrolytes lost in sweat

Muscular system

Involuntary muscle contractions (shivering) work with the skin to control body temperature

Muscles act on facial skin to create expressions

Respiratory system

Stimulation of skin receptors may alter respiratory rate

Nervous system

Sensory receptors provide information about the outside world to the nervous system

Nerves control the activity of sweat glands

Digestive system

Excess calories may be stored as subcutaneous fat

Vitamin D activated by the skin stimulates dietary calcium absorption

Endocrine system

Hormones help to increase skin blood flow during exercise

Other hormones stimulate either the synthesis or the decomposition of subcutaneous fat

Lymphatic system

The skin provides an important first line of defense for the immune system

Cardiovascular system

Skin blood vessels play a role in regulating body temperature

Integumentary System
InnerConnections: Chapter 6

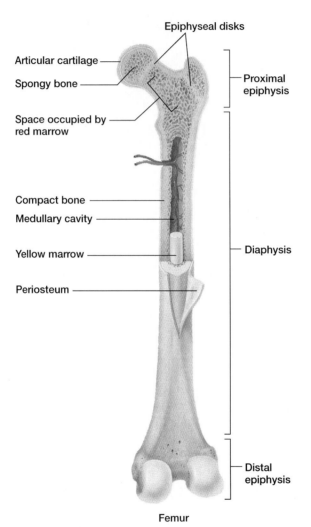

Epiphyseal disks

Articular cartilage

Spongy bone

Space occupied by
red marrow

Compact bone

Medullary cavity

Yellow marrow

Periosteum

Proximal
epiphysis

Diaphysis

Distal
epiphysis

Femur

Structure of a Long Bone
Figure 7.2

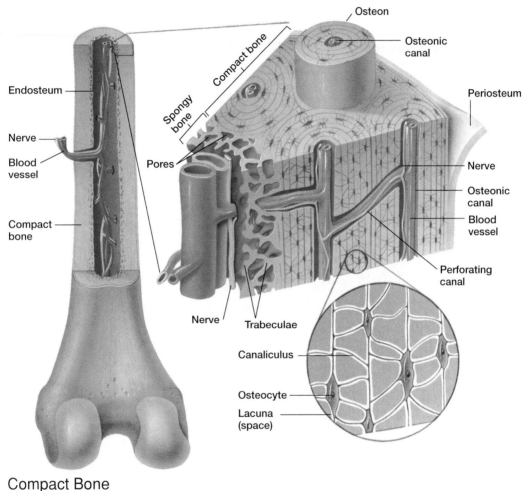

Compact Bone
Figure 7.5

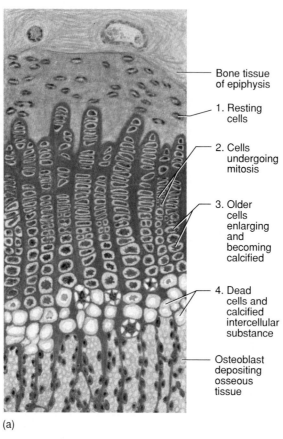

Bone tissue
of epiphysis

1. Resting
cells

2. Cells
undergoing
mitosis

3. Older
cells
enlarging
and
becoming
calcified

4. Dead
cells and
calcified
intercellular
substance

Osteoblast
depositing
osseous
tissue

(a)

Cartilaginous Cells
Figure 7.9a

A *greenstick* fracture is incomplete, and the break occurs on the convex surface of the bend in the bone.

A *fissured* fracture involves an incomplete longitudinal break.

A *comminuted* fracture is complete and fragments the bone.

A *transverse* fracture is complete, and the break occurs at a right angle to the axis of the bone.

An *oblique* fracture occurs at an angle other than a right angle to the axis of the bone.

A *spiral* fracture is caused by twisting a bone excessively.

Traumatic Fractures
Box 7.3,Figure 7A

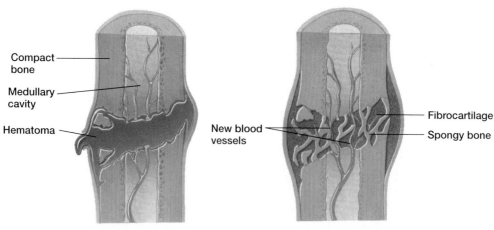

Compact bone

Medullary cavity

Hematoma

New blood vessels

Fibrocartilage

Spongy bone

(a) Blood escapes from ruptured blood vessels and forms a hematoma.

(b) Spongy bone forms in regions close to developing blood vessels, and fibrocartilage forms in more distant regions.

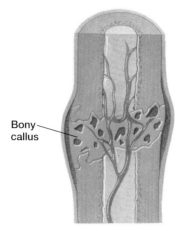

Bony callus

Compact bone

Medullary cavity

Periosteum

(c) A bony callus replaces fibrocartilage.

(d) Osteoclasts remove excess bony tissue, restoring new bone structure much like the original.

Repair of a Fracture
Box 7.3, Figure 7B

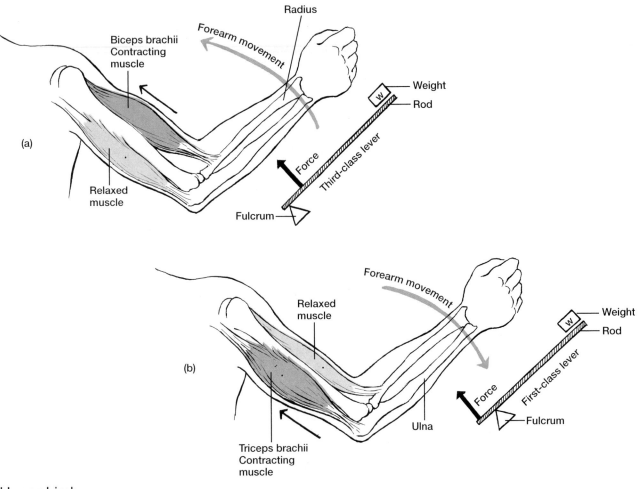

Radius

Biceps brachii
Contracting
muscle

Forearm movement

Weight

Rod

(a)

Force

Third-class lever

Relaxed
muscle

Fulcrum

Forearm movement

Relaxed
muscle

Weight

Rod

(b)

Force

First-class lever

Fulcrum

Ulna

Triceps brachii
Contracting
muscle

Upper Limb
Figure 7.14

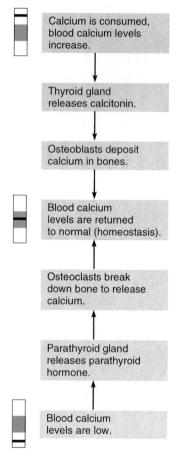

Hormonal Regulation of Bone Calcium
Figure 7.15

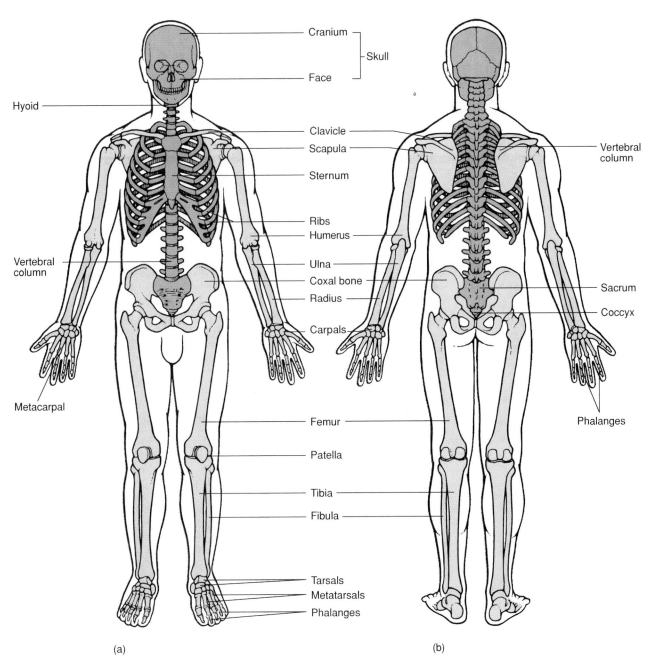

Cranium

Skull

Face

Hyoid

Clavicle

Scapula

Vertebral column

Sternum

Ribs

Humerus

Vertebral column

Ulna

Coxal bone

Radius

Sacrum

Coccyx

Carpals

Metacarpal

Femur

Phalanges

Patella

Tibia

Fibula

Tarsals

Metatarsals

Phalanges

(a)

(b)

Human Skeleton, Anterior & Posterior
Figure 7.17

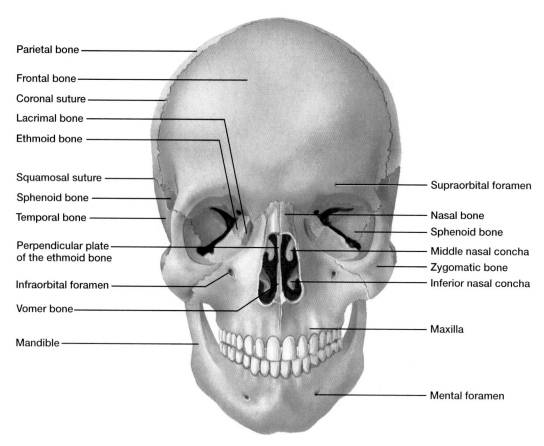

Parietal bone

Frontal bone

Coronal suture

Lacrimal bone

Ethmoid bone

Squamosal suture

Sphenoid bone

Temporal bone

Perpendicular plate
of the ethmoid bone

Infraorbital foramen

Vomer bone

Mandible

Supraorbital foramen

Nasal bone

Sphenoid bone

Middle nasal concha

Zygomatic bone

Inferior nasal concha

Maxilla

Mental foramen

Human Skull, Anterior
Figure 7.19

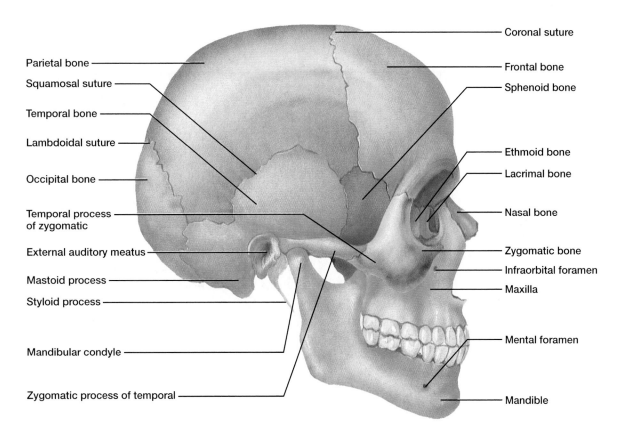

Parietal bone

Squamosal suture

Temporal bone

Lambdoidal suture

Occipital bone

Temporal process
of zygomatic

External auditory meatus

Mastoid process

Styloid process

Mandibular condyle

Zygomatic process of temporal

Coronal suture

Frontal bone

Sphenoid bone

Ethmoid bone

Lacrimal bone

Nasal bone

Zygomatic bone

Infraorbital foramen

Maxilla

Mental foramen

Mandible

Human Skull, Lateral View
Figure 7.21

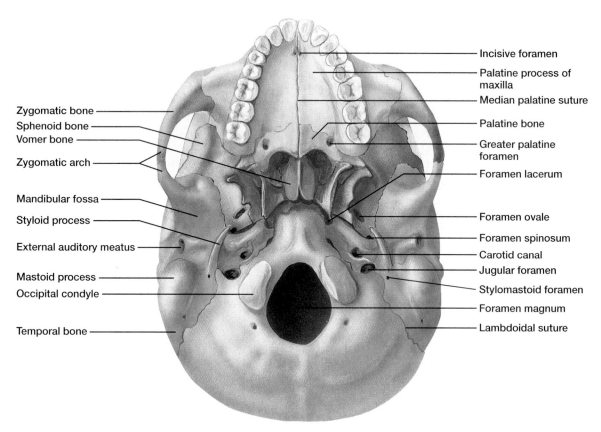

Incisive foramen

Palatine process of maxilla

Median palatine suture

Zygomatic bone

Sphenoid bone

Vomer bone

Zygomatic arch

Palatine bone

Greater palatine foramen

Foramen lacerum

Mandibular fossa

Styloid process

Foramen ovale

External auditory meatus

Foramen spinosum

Carotid canal

Mastoid process

Jugular foramen

Stylomastoid foramen

Occipital condyle

Foramen magnum

Temporal bone

Lambdoidal suture

Human Skull, Inferior
Figure 7.22

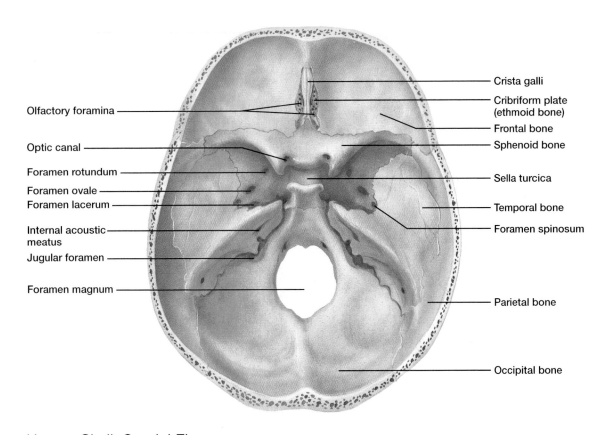

Crista galli

Cribriform plate
(ethmoid bone)

Frontal bone

Sphenoid bone

Olfactory foramina

Optic canal

Foramen rotundum

Sella turcica

Foramen ovale

Foramen lacerum

Internal acoustic
meatus

Temporal bone

Foramen spinosum

Jugular foramen

Foramen magnum

Parietal bone

Occipital bone

Human Skull, Cranial Floor
Figure 7.26

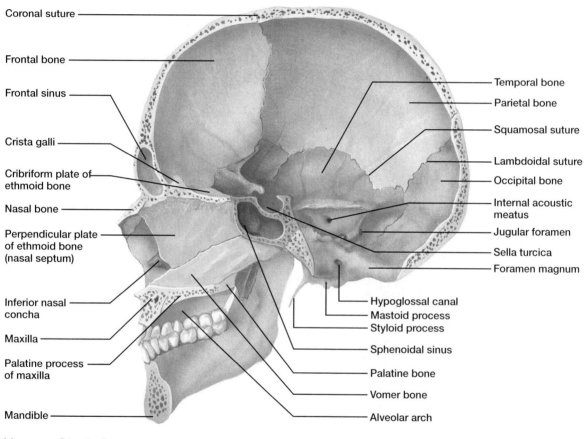

Coronal suture

Frontal bone

Frontal sinus

Crista galli

Cribriform plate of ethmoid bone

Nasal bone

Perpendicular plate of ethmoid bone (nasal septum)

Inferior nasal concha

Maxilla

Palatine process of maxilla

Mandible

Temporal bone

Parietal bone

Squamosal suture

Lambdoidal suture

Occipital bone

Internal acoustic meatus

Jugular foramen

Sella turcica

Foramen magnum

Hypoglossal canal

Mastoid process

Styloid process

Sphenoidal sinus

Palatine bone

Vomer bone

Alveolar arch

Human Skull, Sagittal
Figure 7.29

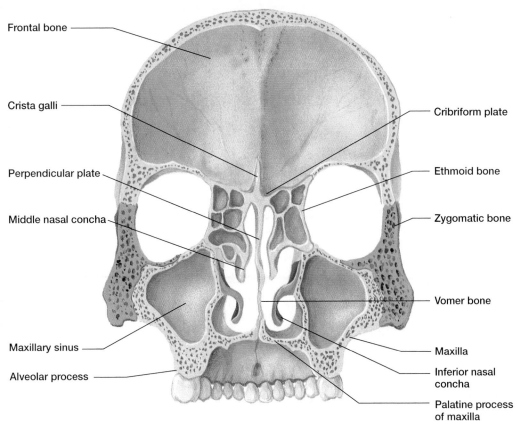

Frontal bone

Crista galli

Perpendicular plate

Middle nasal concha

Maxillary sinus

Alveolar process

Cribriform plate

Ethmoid bone

Zygomatic bone

Vomer bone

Maxilla

Inferior nasal concha

Palatine process of maxilla

Human Skull, Posterior
Figure 7.30

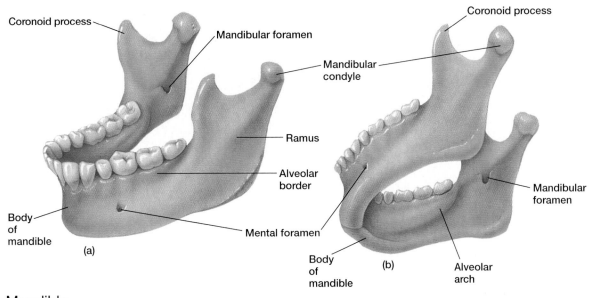

Coronoid process

Mandibular foramen

Mandibular condyle

Coronoid process

Ramus

Alveolar border

Body of mandible

Mental foramen

(a)

Body of mandible

Alveolar arch

Mandibular foramen

(b)

Mandible
Figure 7.31

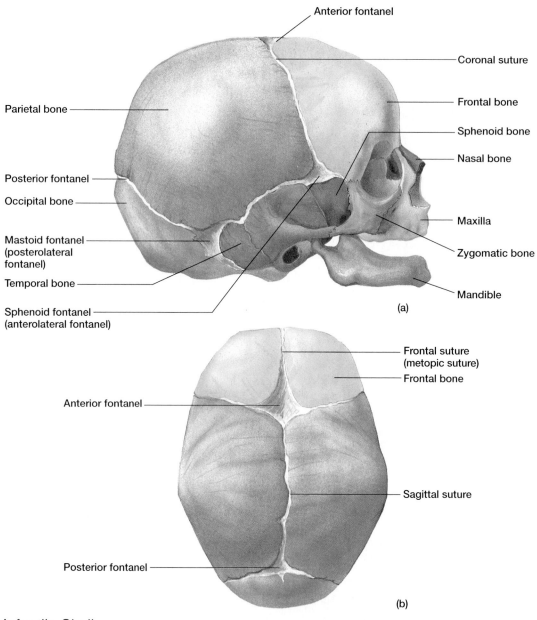

Anterior fontanel

Coronal suture

Frontal bone

Sphenoid bone

Nasal bone

Parietal bone

Posterior fontanel

Occipital bone

Mastoid fontanel
(posterolateral
fontanel)

Temporal bone

Sphenoid fontanel
(anterolateral fontanel)

Maxilla

Zygomatic bone

Mandible

(a)

Frontal suture
(metopic suture)

Frontal bone

Anterior fontanel

Sagittal suture

Posterior fontanel

(b)

Infantile Skull
Figure 7.33

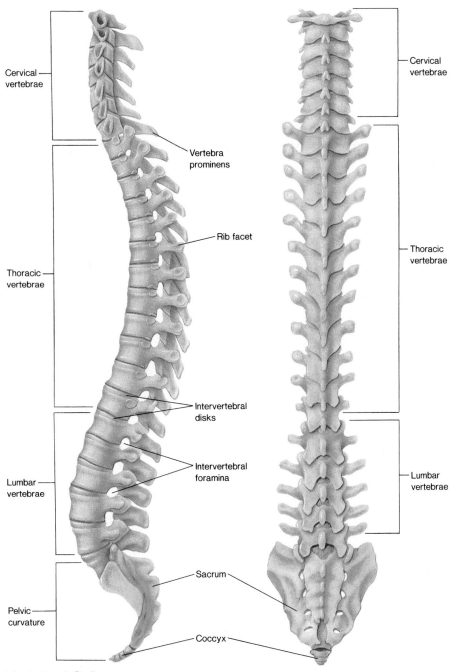

Cervical vertebrae

Vertebra prominens

Rib facet

Thoracic vertebrae

Intervertebral disks

Intervertebral foramina

Lumbar vertebrae

Sacrum

Pelvic curvature

Coccyx

Cervical vertebrae

Thoracic vertebrae

Lumbar vertebrae

Vertebral Column
Figure 7.34

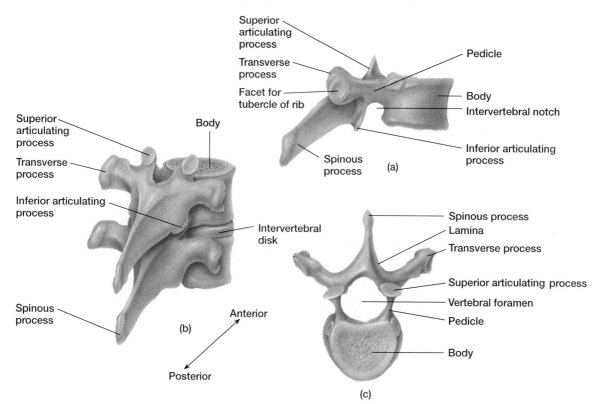

Superior
articulating
process

Transverse
process

Facet for
tubercle of rib

Pedicle

Body

Intervertebral notch

Inferior articulating
process

Spinous
process

(a)

Superior
articulating
process

Transverse
process

Inferior articulating
process

Body

Intervertebral
disk

Spinous
process

(b)

Anterior

Posterior

Spinous process

Lamina

Transverse process

Superior articulating process

Vertebral foramen

Pedicle

Body

(c)

Types of Vertebrae
Figure 7.35

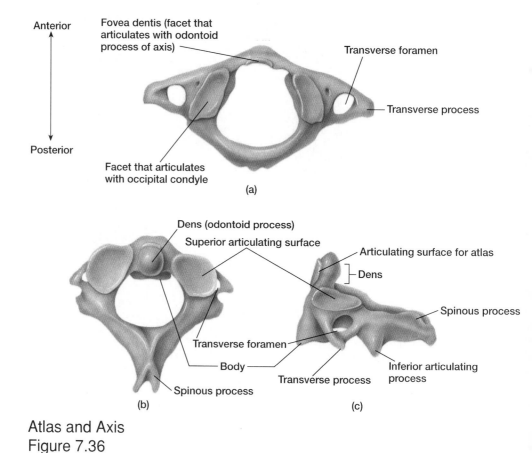

Anterior

Posterior

Fovea dentis (facet that
articulates with odontoid
process of axis)

Transverse foramen

Transverse process

Facet that articulates
with occipital condyle

(a)

Dens (odontoid process)

Superior articulating surface

Articulating surface for atlas

Dens

Spinous process

Transverse foramen

Body

Spinous process

Transverse process

Inferior articulating
process

(b)

(c)

Atlas and Axis
Figure 7.36

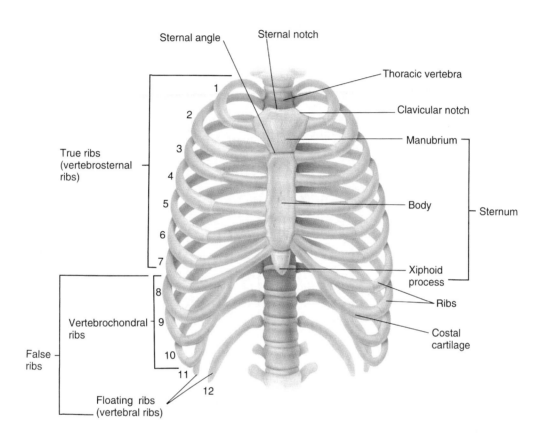

Sternal angle

Sternal notch

Thoracic vertebra

Clavicular notch

Manubrium

Body

Sternum

Xiphoid process

Ribs

Costal cartilage

True ribs (vertebrosternal ribs)

1
2
3
4
5
6
7

Vertebrochondral ribs

8
9

10
11
12

False ribs

Floating ribs (vertebral ribs)

(a)

Thoracic Cage
Figure 7.40a

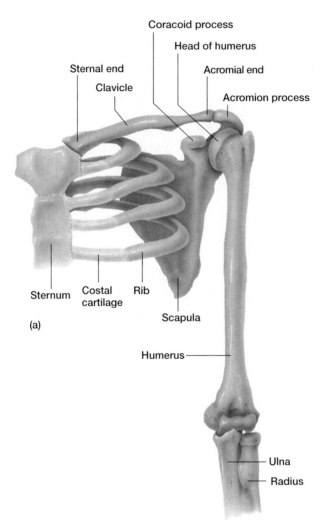

Coracoid process

Head of humerus

Sternal end

Acromial end

Clavicle

Acromion process

Sternum

Costal cartilage

Rib

Scapula

(a)

Humerus

Ulna

Radius

Pectoral Girdle
Figure 7.42a

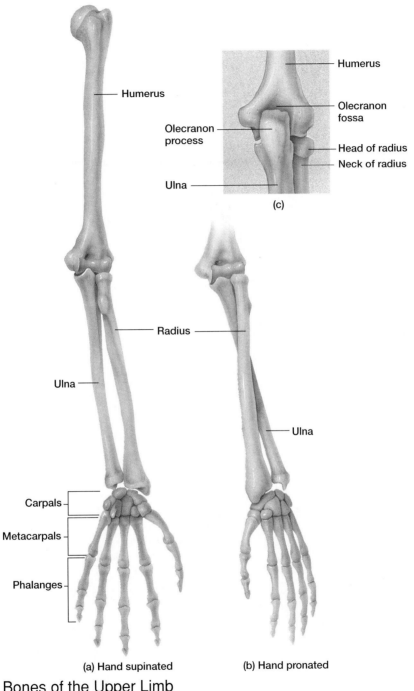

Humerus

Olecranon
process

Ulna

Humerus

Olecranon
fossa

Head of radius
Neck of radius

(c)

Radius

Ulna

Ulna

Carpals

Metacarpals

Phalanges

(a) Hand supinated

(b) Hand pronated

Bones of the Upper Limb
Figure 7.44a

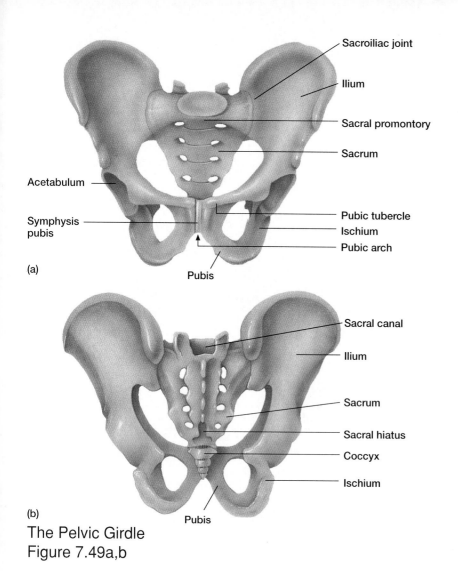

(a)

(b)

The Pelvic Girdle
Figure 7.49a,b

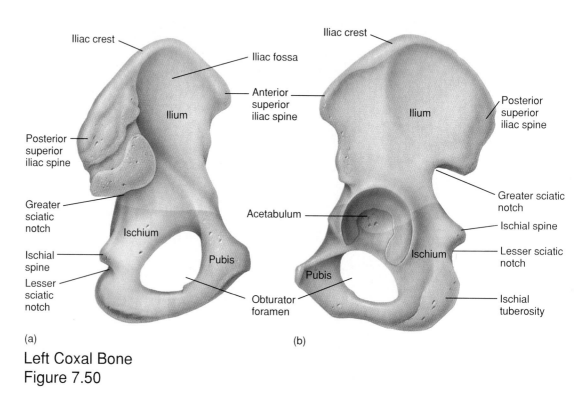

(a)

(b)

Left Coxal Bone
Figure 7.50

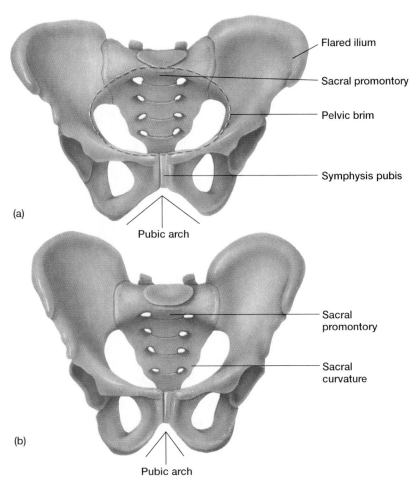

(a)

Flared ilium

Sacral promontory

Pelvic brim

Symphysis pubis

Pubic arch

(b)

Sacral promontory

Sacral curvature

Pubic arch

Female Pelvis
Figure 7.51

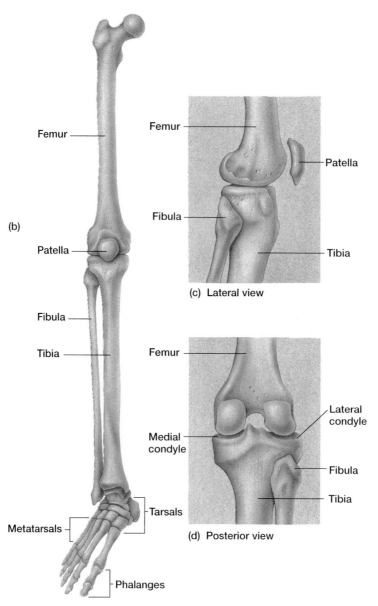

(b)

Femur

Patella

Fibula

Tibia

Metatarsals

Tarsals

Phalanges

(c) Lateral view

Femur

Patella

Fibula

Tibia

(d) Posterior view

Femur

Medial condyle

Lateral condyle

Fibula

Tibia

Lower Right Limb
Figure 7.52b,c,d

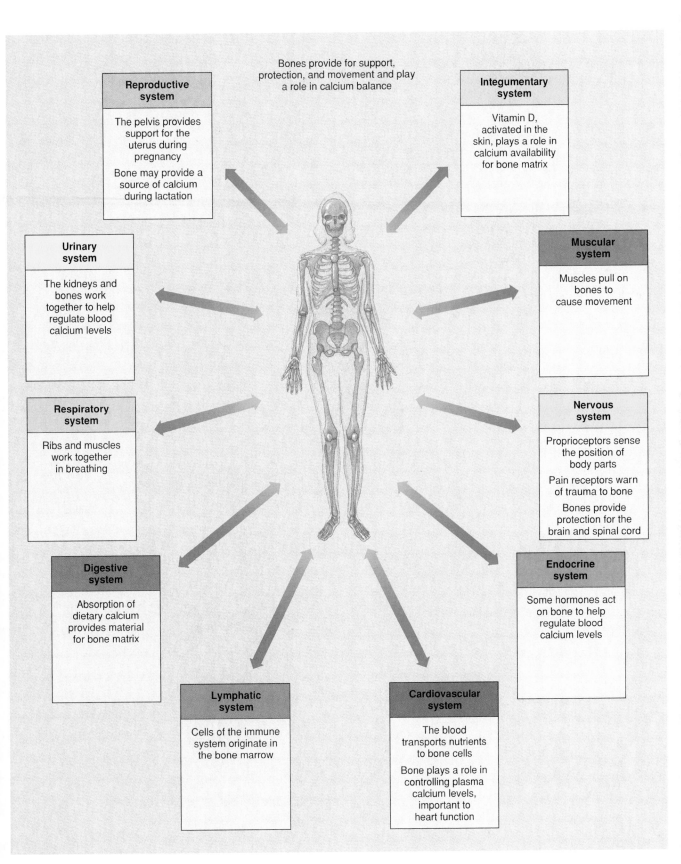

Bones provide for support, protection, and movement and play a role in calcium balance

Reproductive system

The pelvis provides support for the uterus during pregnancy

Bone may provide a source of calcium during lactation

Integumentary system

Vitamin D, activated in the skin, plays a role in calcium availability for bone matrix

Urinary system

The kidneys and bones work together to help regulate blood calcium levels

Muscular system

Muscles pull on bones to cause movement

Respiratory system

Ribs and muscles work together in breathing

Nervous system

Proprioceptors sense the position of body parts

Pain receptors warn of trauma to bone

Bones provide protection for the brain and spinal cord

Digestive system

Absorption of dietary calcium provides material for bone matrix

Endocrine system

Some hormones act on bone to help regulate blood calcium levels

Lymphatic system

Cells of the immune system originate in the bone marrow

Cardiovascular system

The blood transports nutrients to bone cells

Bone plays a role in controlling plasma calcium levels, important to heart function

Skeletal System
InnerConnections: Chapter 7

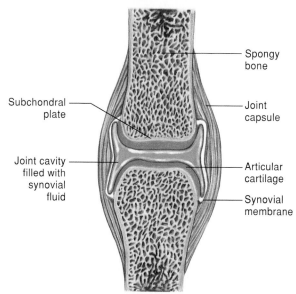

Spongy
bone

Subchondral
plate

Joint
capsule

Joint cavity
filled with
synovial
fluid

Articular
cartilage

Synovial
membrane

Synovial Joint
Figure 8.7

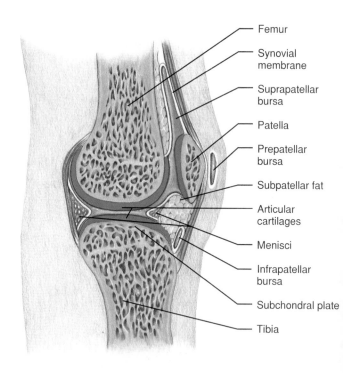

Femur

Synovial
membrane

Suprapatellar
bursa

Patella

Prepatellar
bursa

Subpatellar fat

Articular
cartilages

Menisci

Infrapatellar
bursa

Subchondral plate

Tibia

Femur and Tibia
Figure 8.8

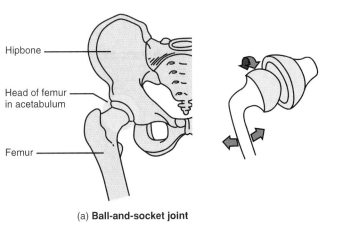

Hipbone

Head of femur
in acetabulum

Femur

(a) **Ball-and-socket joint**

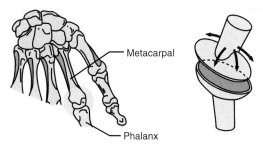

Metacarpal

Phalanx

(b) **Condyloid joint**

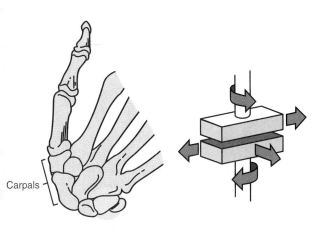

Carpals

(c) **Gliding joint**

Movable Joints
Figure 8.9a-c

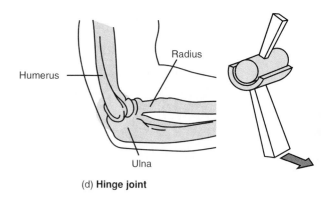

(d) **Hinge joint**

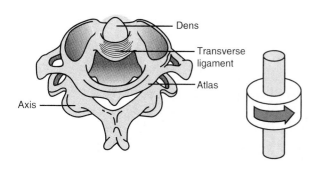

(e) **Pivot joint**

Movable Joints
Figure 8.9d-f

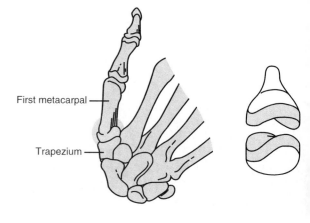

(f) **Saddle joint**

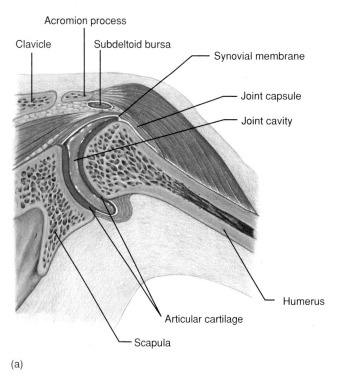

(a)

Shoulder Joint
Figure 8.13a

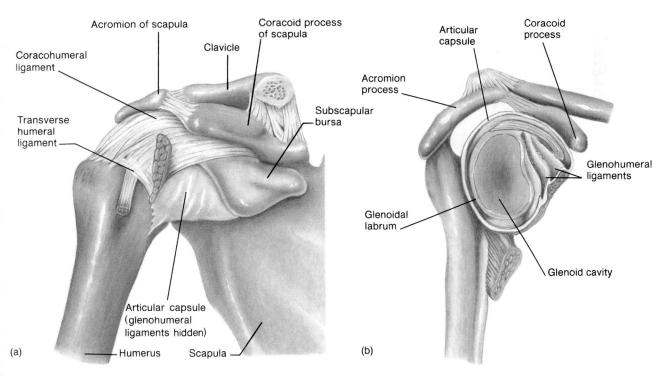

Surfaces of the Shoulder
Figure 8.14

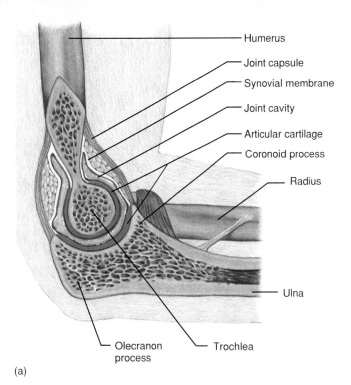

(a)

Elbow Joint
Figure 8.15a

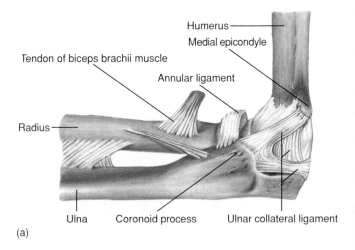

(a)

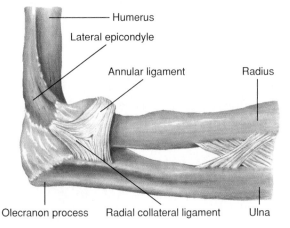

(b)

Collateral Ligament
Figure 8.16

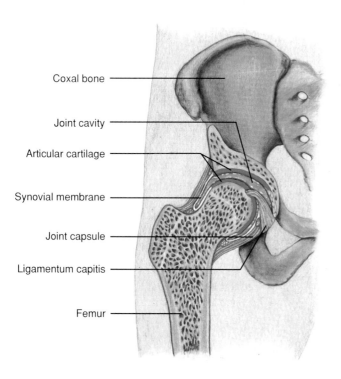

Coxal bone

Joint cavity

Articular cartilage

Synovial membrane

Joint capsule

Ligamentum capitis

Femur

(a)

Hip Joint
Figure 8.18a

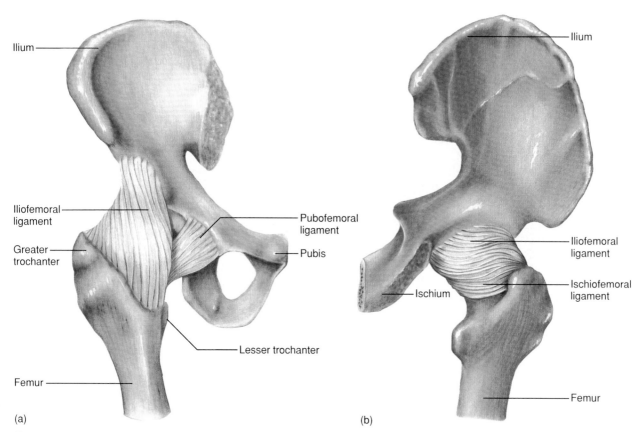

Ilium

Iliofemoral ligament

Greater trochanter

Femur

(a)

Pubofemoral ligament

Pubis

Lesser trochanter

Ilium

Iliofemoral ligament

Ischiofemoral ligament

Ischium

Femur

(b)

Right Hip Joint
Figure 8.19

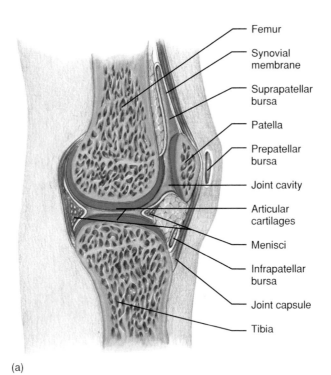

Femur

Synovial membrane

Suprapatellar bursa

Patella

Prepatellar bursa

Joint cavity

Articular cartilages

Menisci

Infrapatellar bursa

Joint capsule

Tibia

(a)

Knee Joint
Figure 8.20a

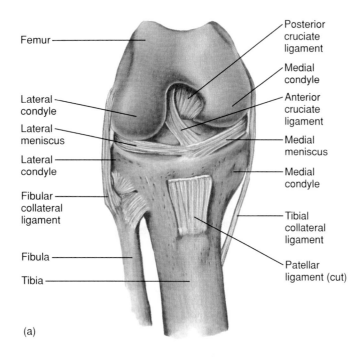

Femur

Posterior cruciate ligament

Medial condyle

Lateral condyle

Anterior cruciate ligament

Lateral meniscus

Medial meniscus

Lateral condyle

Medial condyle

Fibular collateral ligament

Tibial collateral ligament

Fibula

Tibia

Patellar ligament (cut)

(a)

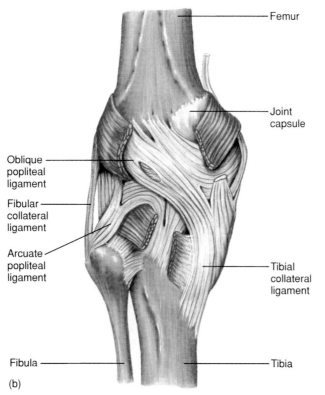

Femur

Joint capsule

Oblique popliteal ligament

Fibular collateral ligament

Arcuate popliteal ligament

Tibial collateral ligament

Fibula

Tibia

(b)

Ligaments within the Knee Joint
Figure 8.21

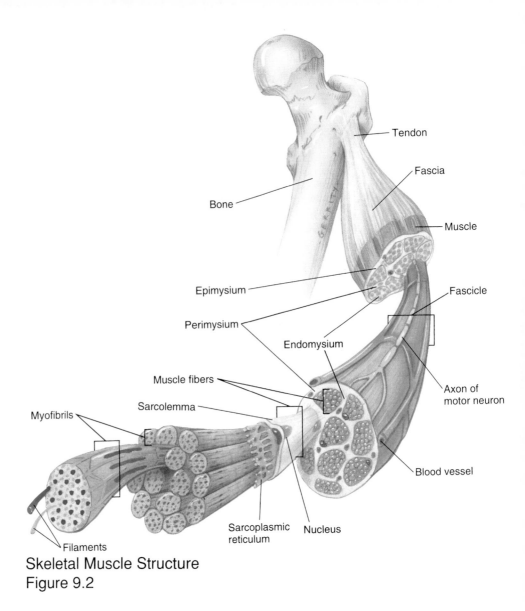

Tendon

Fascia

Bone

Muscle

Epimysium

Perimysium

Fascicle

Endomysium

Axon of
motor neuron

Muscle fibers

Sarcolemma

Myofibrils

Blood vessel

Sarcoplasmic
reticulum

Nucleus

Filaments

Skeletal Muscle Structure
Figure 9.2

Fasciculus

Epimysium

Endomysium

Muscle fiber

Perimysium

Skeletal Muscle Fiber I
Figure 9.3

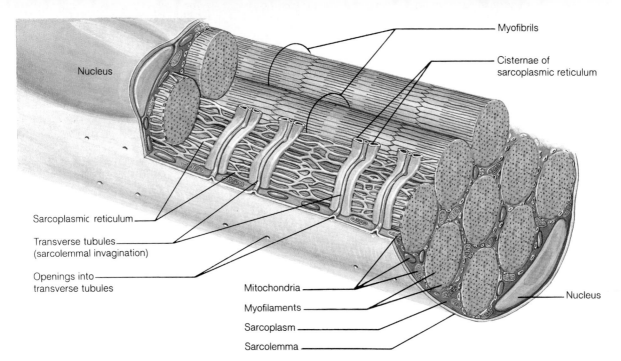

Nucleus

Myofibrils

Cisternae of
sarcoplasmic reticulum

Sarcoplasmic reticulum

Transverse tubules
(sarcolemmal invagination)

Openings into
transverse tubules

Mitochondria

Myofilaments

Sarcoplasm

Sarcolemma

Nucleus

Skeletal Muscle Fiber II
Figure 9.6

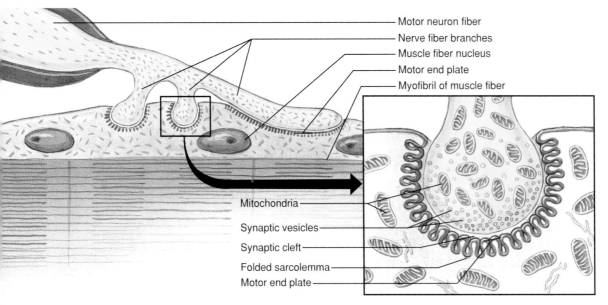

Motor neuron fiber

Nerve fiber branches

Muscle fiber nucleus

Motor end plate

Myofibril of muscle fiber

Mitochondria

Synaptic vesicles

Synaptic cleft

Folded sarcolemma

Motor end plate

Neuromuscular Junction
Figure 9.7

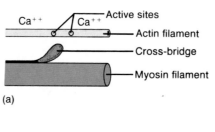

(a)

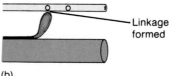

(b)

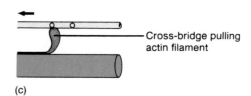

(c)

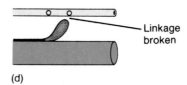

(d)

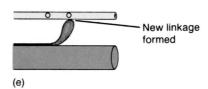

(e)

Sliding Filament Theory
Figure 9.10

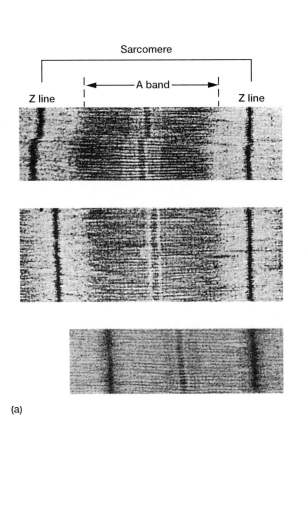

(a)

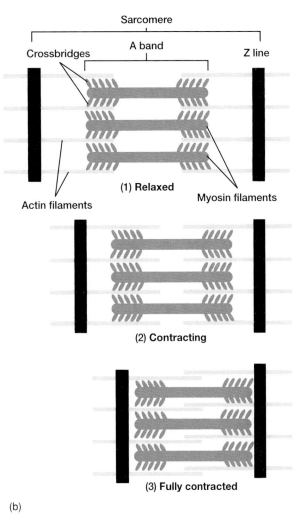

Sarcomere

A band

Crossbridges

Z line

Actin filaments

(1) Relaxed

Myosin filaments

(2) Contracting

(3) Fully contracted

(b)

Contraction of a Sarcomere
Figure 9.11

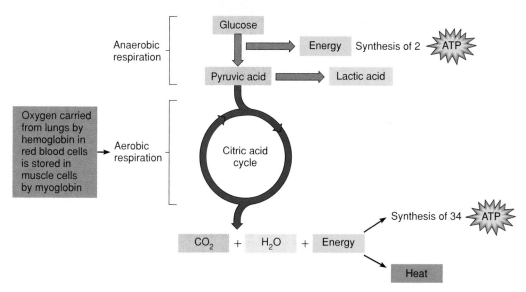

Muscle Metabolism
Figure 9.13

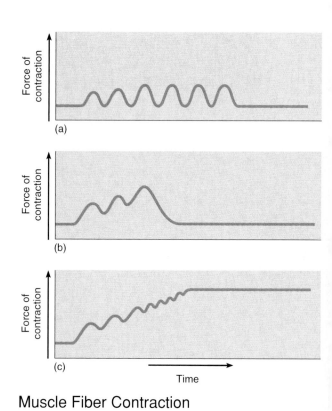

Muscle Fiber Contraction
Figure 9.16

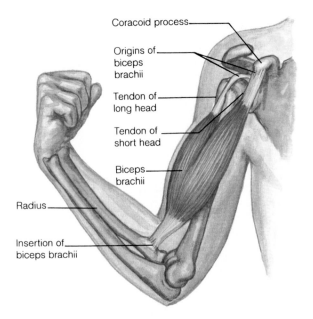

Coracoid process

Origins of
biceps
brachii

Tendon of
long head

Tendon of
short head

Biceps
brachii

Radius

Insertion of
biceps brachii

Origin and Insertion
Figure 9.19

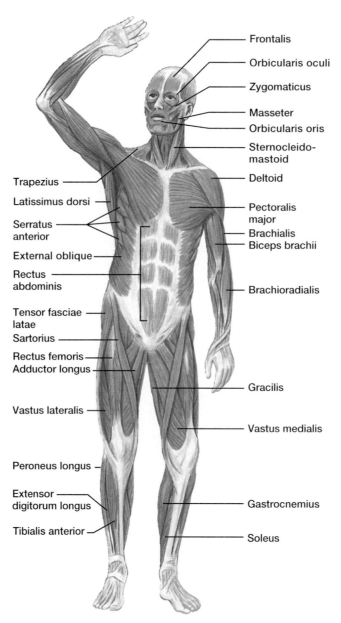

Frontalis

Orbicularis oculi

Zygomaticus

Masseter

Orbicularis oris

Sternocleido-
mastoid

Deltoid

Pectoralis
major

Brachialis

Biceps brachii

Brachioradialis

Gracilis

Vastus medialis

Gastrocnemius

Soleus

Trapezius

Latissimus dorsi

Serratus
anterior

External oblique

Rectus
abdominis

Tensor fasciae
latae

Sartorius

Rectus femoris

Adductor longus

Vastus lateralis

Peroneus longus

Extensor
digitorum longus

Tibialis anterior

Skeletal Muscles, Anterior
Figure 9.20

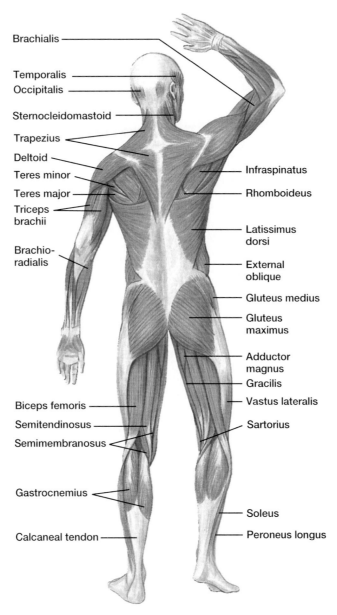

Brachialis

Temporalis

Occipitalis

Sternocleidomastoid

Trapezius

Deltoid

Teres minor

Teres major

Triceps
brachii

Brachio-
radialis

Infraspinatus

Rhomboideus

Latissimus
dorsi

External
oblique

Gluteus medius

Gluteus
maximus

Adductor
magnus

Gracilis

Vastus lateralis

Biceps femoris

Semitendinosus

Semimembranosus

Sartorius

Gastrocnemius

Soleus

Peroneus longus

Calcaneal tendon

Skeletal Muscles, Posterior
Figure 9.21

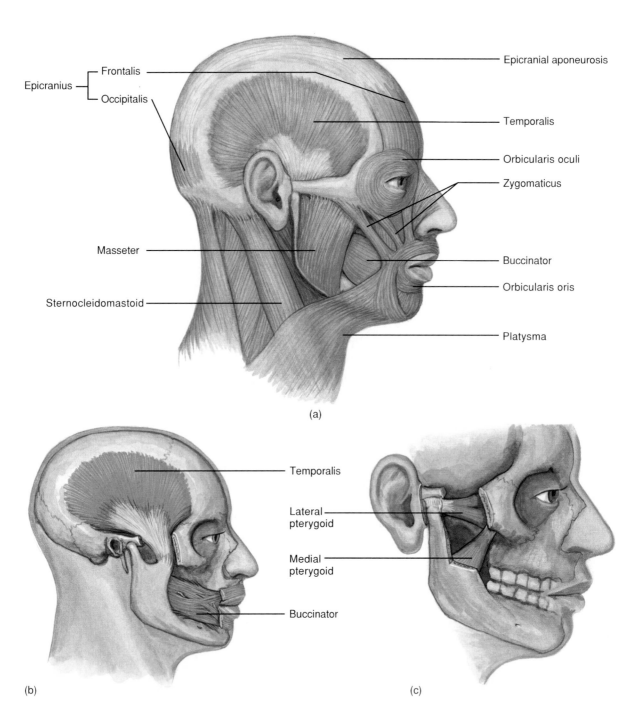

Epicranius — Frontalis
— Occipitalis

Epicranial aponeurosis

Temporalis

Orbicularis oculi

Zygomaticus

Masseter

Buccinator

Orbicularis oris

Sternocleidomastoid

Platysma

(a)

Temporalis

Lateral pterygoid

Medial pterygoid

Buccinator

(b)

(c)

Muscles of Expression and Mastication
Figure 9.22

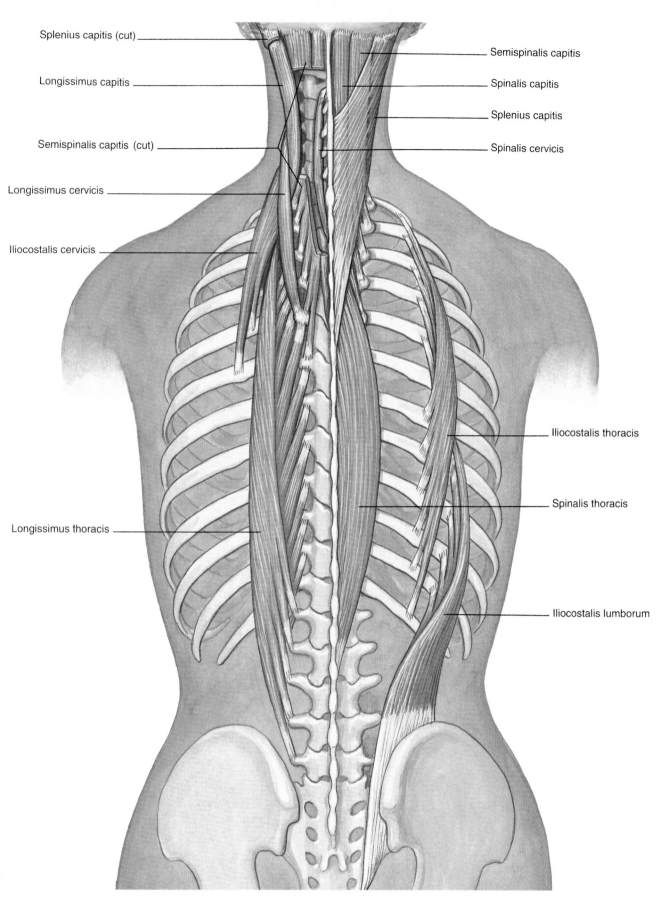

Splenius capitis (cut)

Longissimus capitis

Semispinalis capitis (cut)

Longissimus cervicis

Iliocostalis cervicis

Longissimus thoracis

Semispinalis capitis

Spinalis capitis

Splenius capitis

Spinalis cervicis

Iliocostalis thoracis

Spinalis thoracis

Iliocostalis lumborum

Muscles of Posterior Neck and Trunk
Figure 9.23

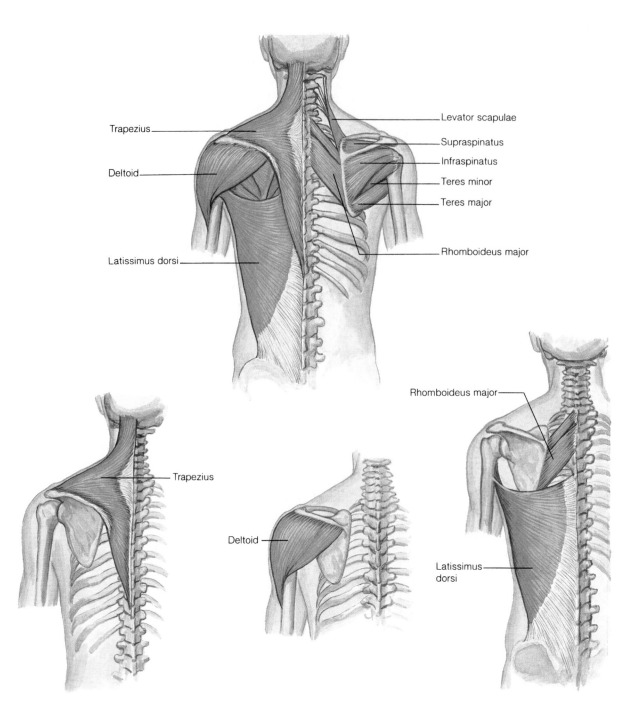

Muscles of Posterior Shoulder
Figure 9.24

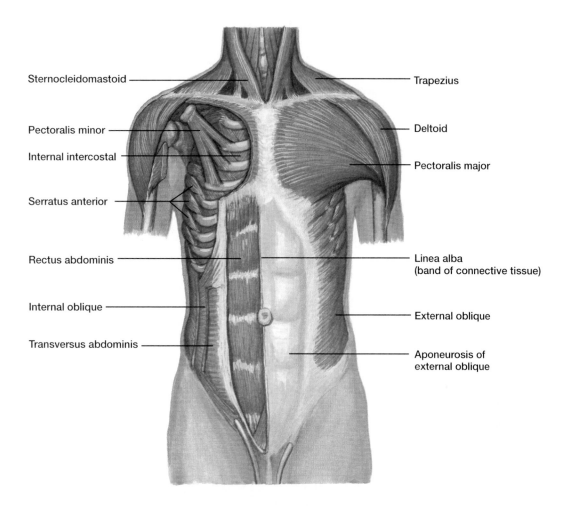

Sternocleidomastoid

Pectoralis minor

Internal intercostal

Serratus anterior

Rectus abdominis

Internal oblique

Transversus abdominis

Trapezius

Deltoid

Pectoralis major

Linea alba
(band of connective tissue)

External oblique

Aponeurosis of
external oblique

Muscles of the Anterior Chest and Abdominal Wall
Figure 9.25

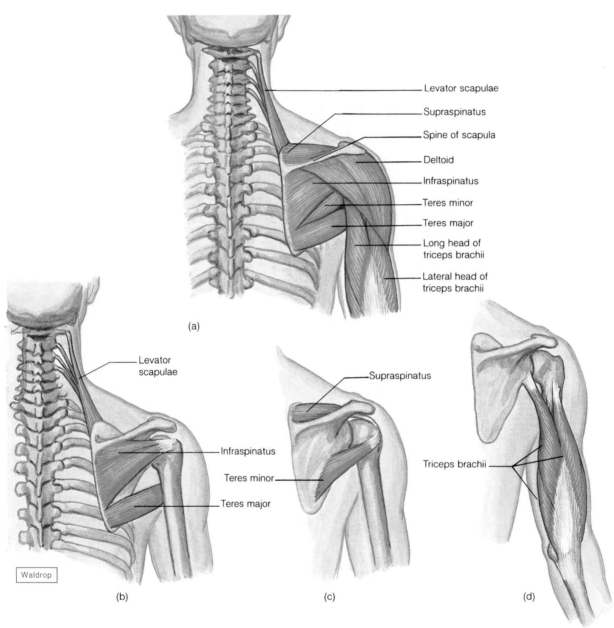

(a)

Levator scapulae
Supraspinatus
Spine of scapula
Deltoid
Infraspinatus
Teres minor
Teres major
Long head of
triceps brachii
Lateral head of
triceps brachii

Levator
scapulae

Infraspinatus

Teres minor

Teres major

Waldrop

(b)

Supraspinatus

Teres minor

(c)

Triceps brachii

(d)

Muscles of the Scapula and Upper Arm
Figure 9.26

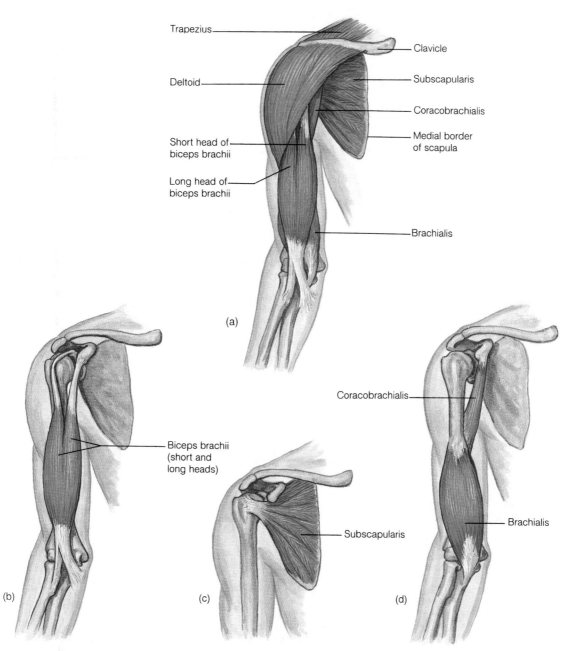

Muscles of the Anterior Shoulder and Upper Arm
Figure 9.28

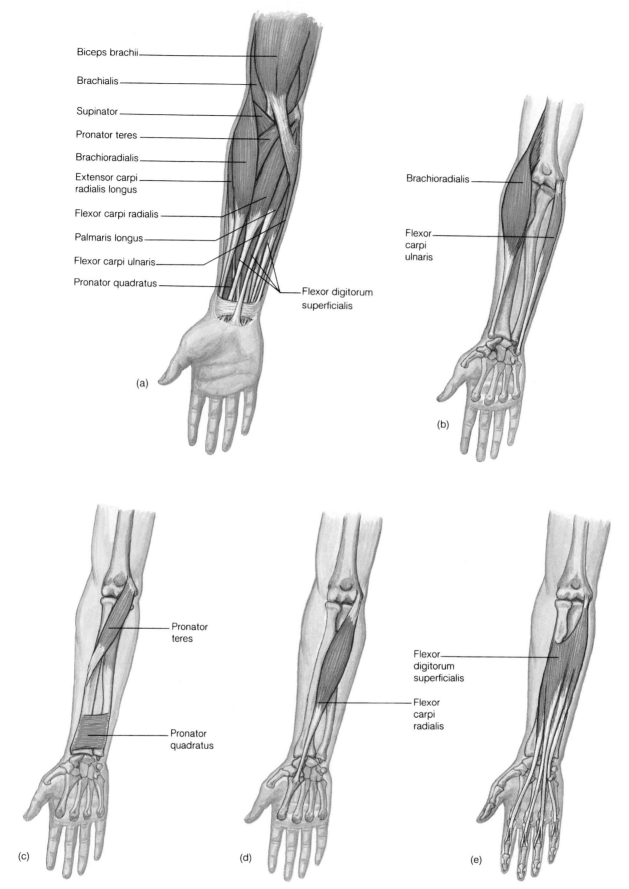

Biceps brachii

Brachialis

Supinator

Pronator teres

Brachioradialis

Extensor carpi radialis longus

Flexor carpi radialis

Palmaris longus

Flexor carpi ulnaris

Pronator quadratus

Flexor digitorum superficialis

(a)

Brachioradialis

Flexor carpi ulnaris

(b)

Pronator teres

Pronator quadratus

(c)

Flexor digitorum superficialis

Flexor carpi radialis

(d)

(e)

Muscles of the Anterior Forearm
Figure 9.29

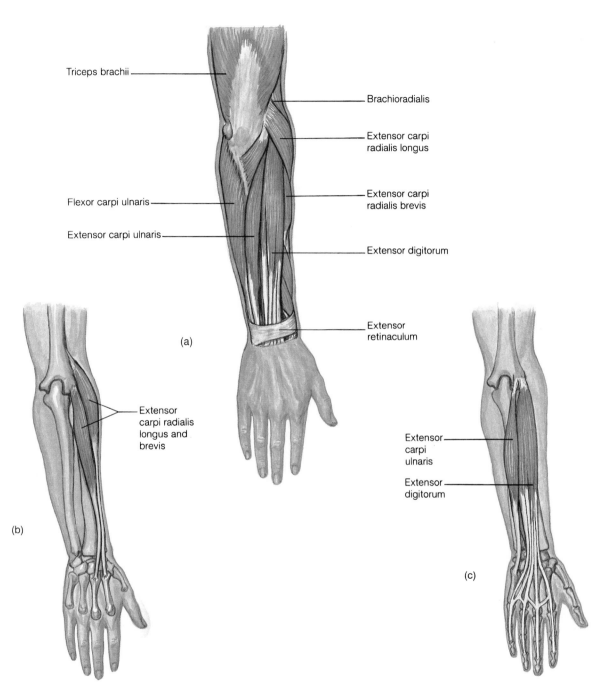

Triceps brachii

Brachioradialis

Extensor carpi radialis longus

Flexor carpi ulnaris

Extensor carpi radialis brevis

Extensor carpi ulnaris

Extensor digitorum

Extensor retinaculum

(a)

Extensor carpi radialis longus and brevis

(b)

Extensor carpi ulnaris

Extensor digitorum

(c)

Muscles of the Posterior Forearm
Figure 9.30

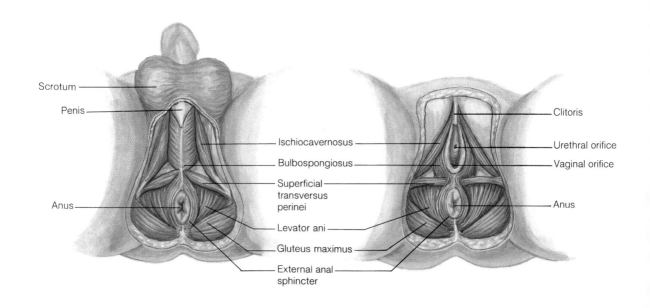

Scrotum

Penis

Anus

Ischiocavernosus

Bulbospongiosus

Superficial
transversus
perinei

Levator ani

Gluteus maximus

External anal
sphincter

Clitoris

Urethral orifice

Vaginal orifice

Anus

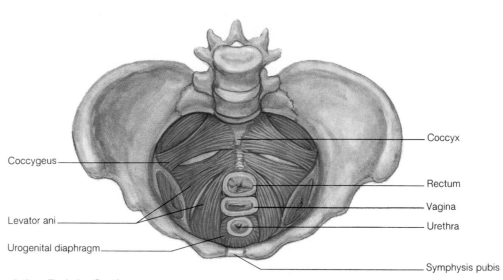

Coccygeus

Levator ani

Urogenital diaphragm

Coccyx

Rectum

Vagina

Urethra

Symphysis pubis

Muscles of the Pelvic Outlet
Figure 9.33

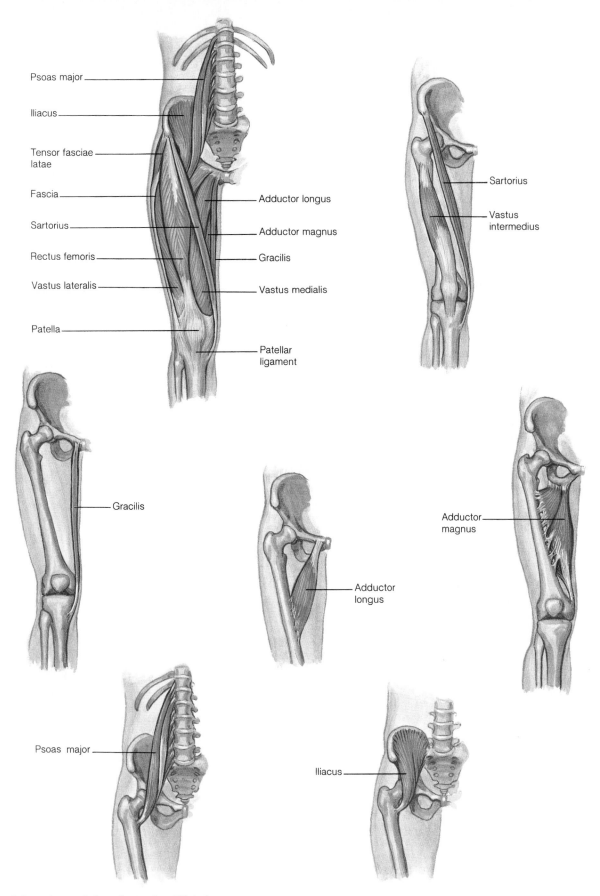

Psoas major

Iliacus

Tensor fasciae latae

Fascia

Sartorius

Rectus femoris

Vastus lateralis

Patella

Adductor longus

Adductor magnus

Gracilis

Vastus medialis

Patellar ligament

Sartorius

Vastus intermedius

Gracilis

Adductor longus

Adductor magnus

Psoas major

Iliacus

Muscles of the Anterior Thigh
Figure 9.34

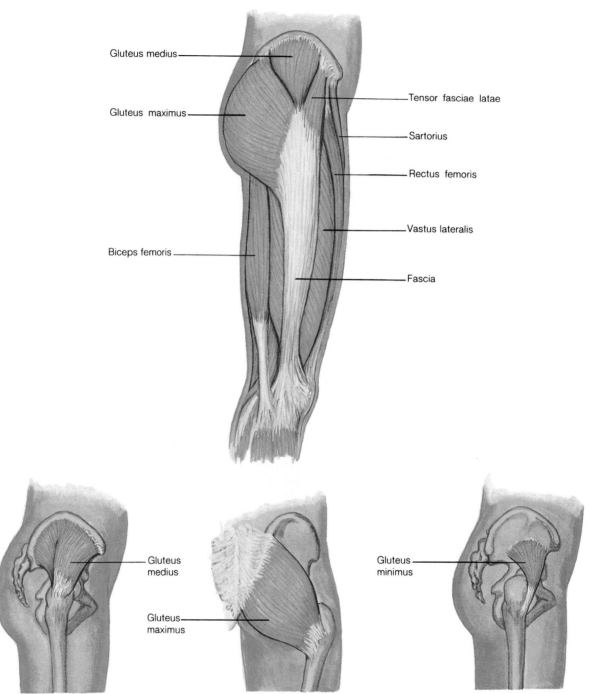

Gluteus medius

Gluteus maximus

Biceps femoris

Tensor fasciae latae

Sartorius

Rectus femoris

Vastus lateralis

Fascia

Gluteus medius

Gluteus maximus

Gluteus minimus

Muscles of the Lateral Thigh
Figure 9.35

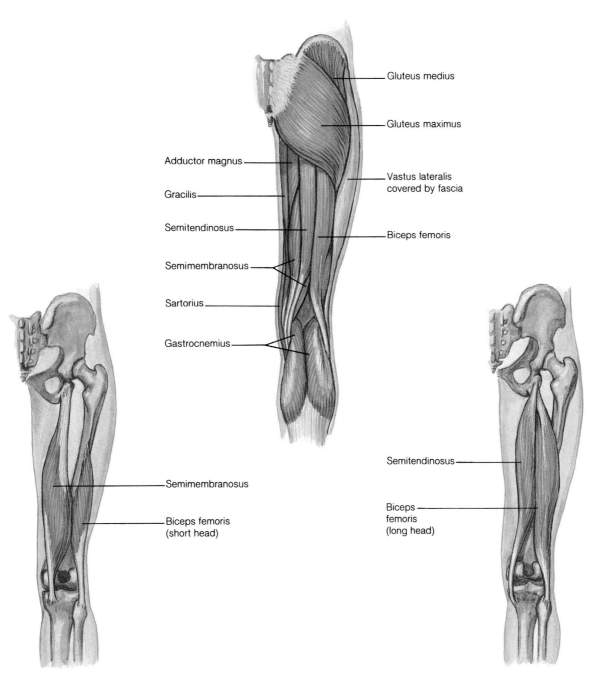

Gluteus medius

Gluteus maximus

Adductor magnus

Vastus lateralis
covered by fascia

Gracilis

Semitendinosus

Biceps femoris

Semimembranosus

Sartorius

Gastrocnemius

Semimembranosus

Semitendinosus

Biceps femoris
(short head)

Biceps
femoris
(long head)

Muscles of the Posterior Thigh
Figure 9.36

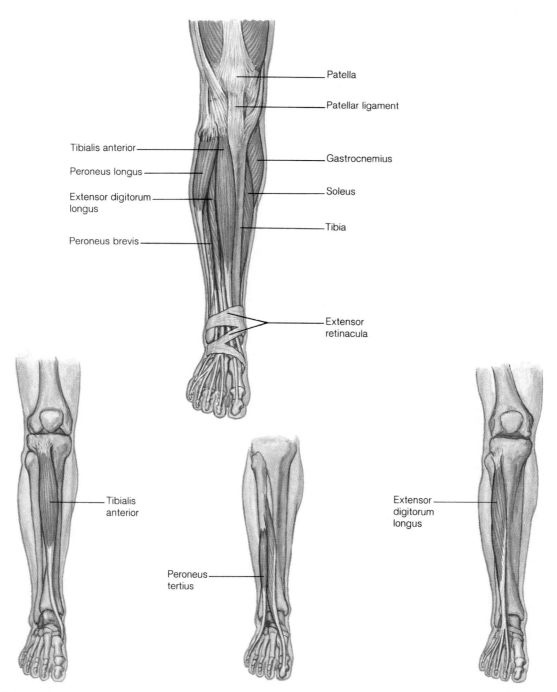

Patella

Patellar ligament

Tibialis anterior

Peroneus longus

Extensor digitorum
longus

Peroneus brevis

Gastrocnemius

Soleus

Tibia

Extensor
retinacula

Tibialis
anterior

Peroneus
tertius

Extensor
digitorum
longus

Muscles of the Anterior Lower Leg
Figure 9.38

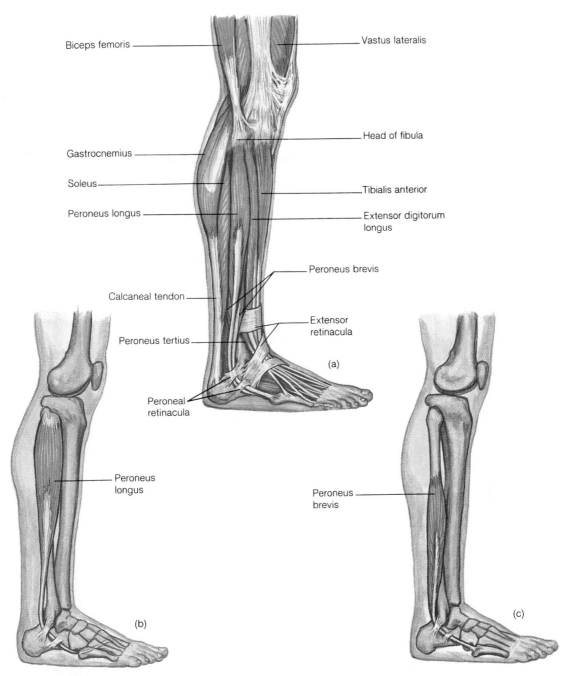

Biceps femoris

Vastus lateralis

Head of fibula

Gastrocnemius

Soleus

Tibialis anterior

Peroneus longus

Extensor digitorum longus

Peroneus brevis

Calcaneal tendon

Extensor retinacula

Peroneus tertius

(a)

Peroneal retinacula

Peroneus longus

Peroneus brevis

(b)

(c)

Muscles of the Lateral Lower Leg
Figure 9.39

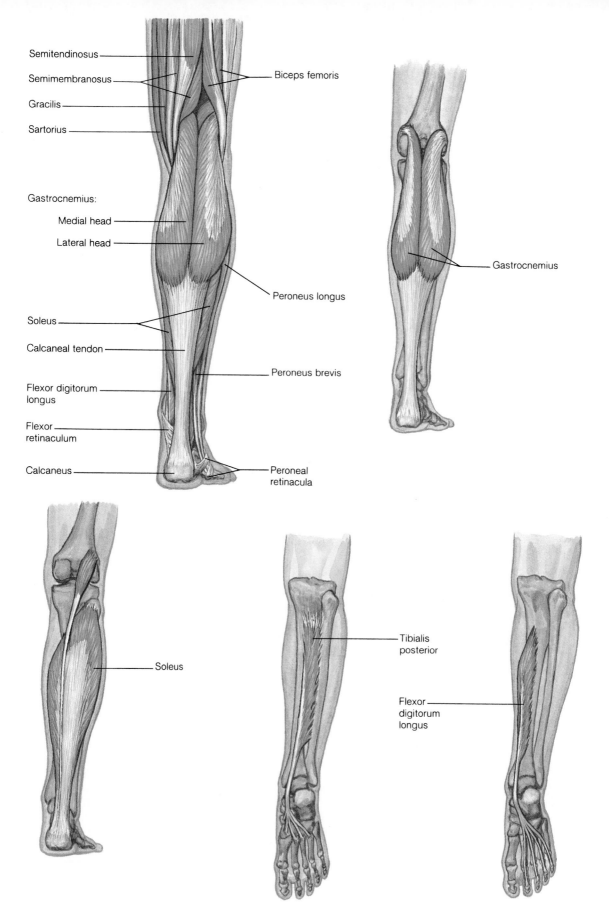

Semitendinosus

Semimembranosus

Gracilis

Sartorius

Biceps femoris

Gastrocnemius:

    Medial head

    Lateral head

Peroneus longus

Soleus

Calcaneal tendon

Flexor digitorum longus

Flexor retinaculum

Calcaneus

Peroneus brevis

Peroneal retinacula

Gastrocnemius

Soleus

Tibialis posterior

Flexor digitorum longus

Muscles of the Posterior Lower Leg
Figure 9.40

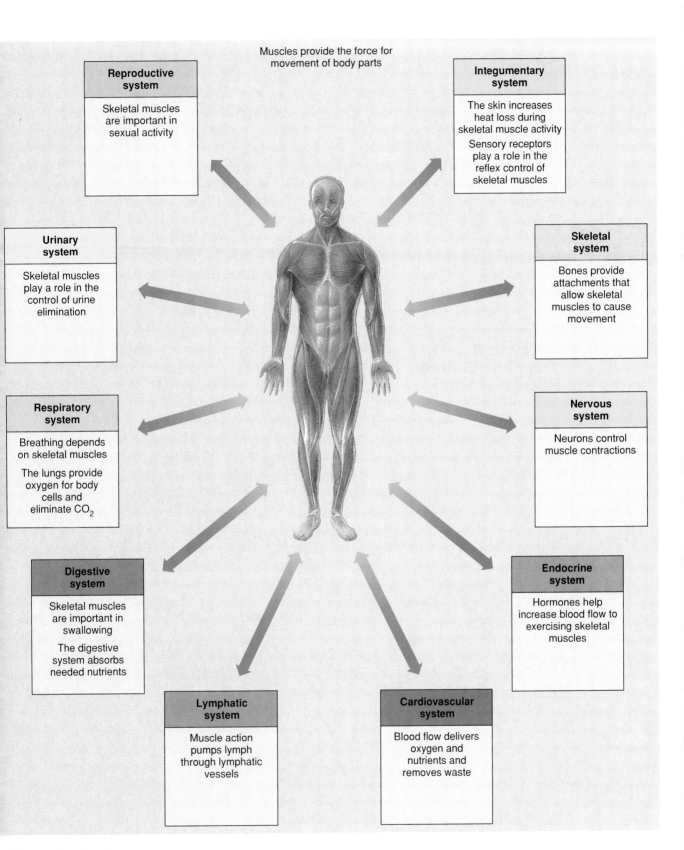

Muscles provide the force for movement of body parts

Reproductive system

Skeletal muscles are important in sexual activity

Integumentary system

The skin increases heat loss during skeletal muscle activity

Sensory receptors play a role in the reflex control of skeletal muscles

Urinary system

Skeletal muscles play a role in the control of urine elimination

Skeletal system

Bones provide attachments that allow skeletal muscles to cause movement

Respiratory system

Breathing depends on skeletal muscles

The lungs provide oxygen for body cells and eliminate CO_2

Nervous system

Neurons control muscle contractions

Digestive system

Skeletal muscles are important in swallowing

The digestive system absorbs needed nutrients

Endocrine system

Hormones help increase blood flow to exercising skeletal muscles

Lymphatic system

Muscle action pumps lymph through lymphatic vessels

Cardiovascular system

Blood flow delivers oxygen and nutrients and removes waste

Muscular System
InnerConnections: Chapter 9

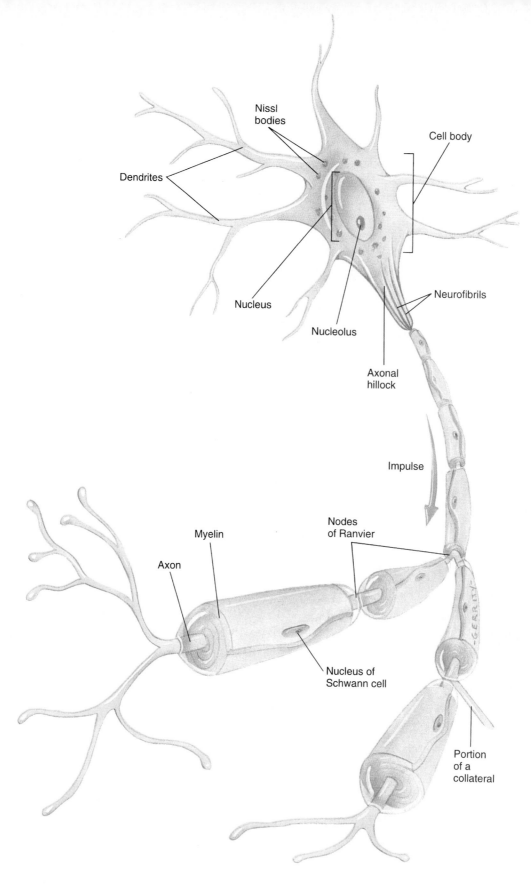

Nissl
bodies

Cell body

Dendrites

Nucleus

Nucleolus

Neurofibrils

Axonal
hillock

Impulse

Nodes
of Ranvier

Myelin

Axon

Nucleus of
Schwann cell

Portion
of a
collateral

Motor and Sensory Neuron
Figure 10.3

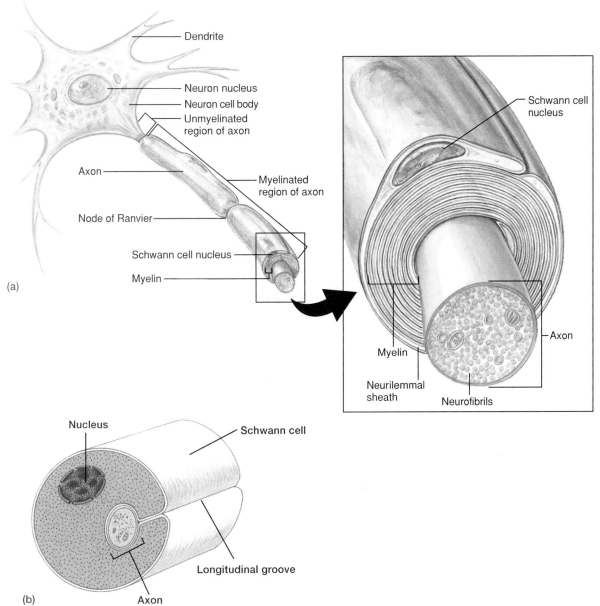

(a)

Dendrite

Neuron nucleus

Neuron cell body

Unmyelinated region of axon

Axon

Myelinated region of axon

Node of Ranvier

Schwann cell nucleus

Myelin

Schwann cell nucleus

Myelin

Neurilemmal sheath

Neurofibrils

Axon

(b)

Nucleus

Schwann cell

Longitudinal groove

Axon

Schwann Cell
Figure 10.4

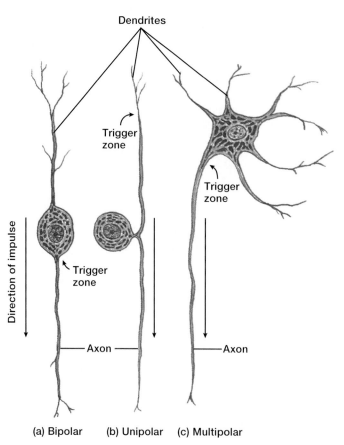

Dendrites

Trigger zone

Trigger zone

Direction of impulse

Trigger zone

Axon

Axon

(a) Bipolar   (b) Unipolar   (c) Multipolar

Types of Neurons
Figure 10.6

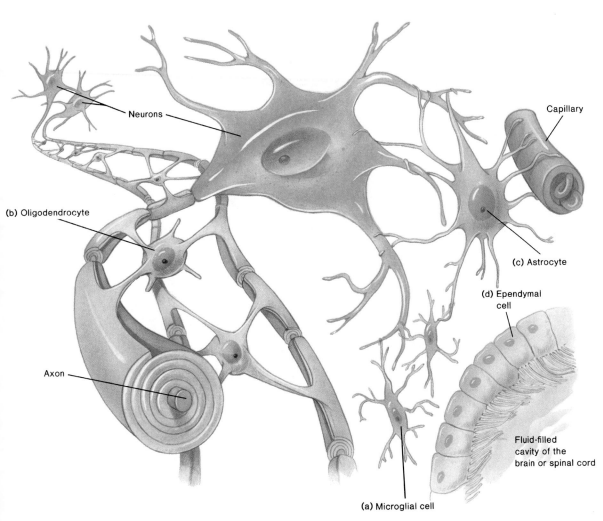

Neurons

Capillary

(b) Oligodendrocyte

(c) Astrocyte

(d) Ependymal cell

Axon

Fluid-filled cavity of the brain or spinal cord

(a) Microglial cell

Neuroglial Cells
Figure 10.7

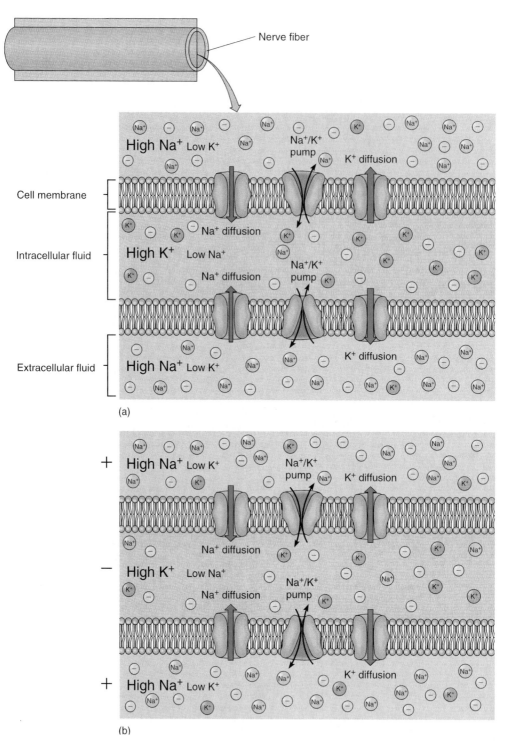

Nerve fiber

Cell membrane

Intracellular fluid

Extracellular fluid

High Na⁺ Low K⁺

Na⁺/K⁺ pump

K⁺ diffusion

Na⁺ diffusion

High K⁺ Low Na⁺

Na⁺ diffusion

Na⁺/K⁺ pump

High Na⁺ Low K⁺

K⁺ diffusion

(a)

High Na⁺ Low K⁺

Na⁺/K⁺ pump

K⁺ diffusion

Na⁺ diffusion

High K⁺ Low Na⁺

Na⁺ diffusion

Na⁺/K⁺ pump

High Na⁺ Low K⁺

K⁺ diffusion

(b)

Establishment of the Resting Potential
Figure 10.11

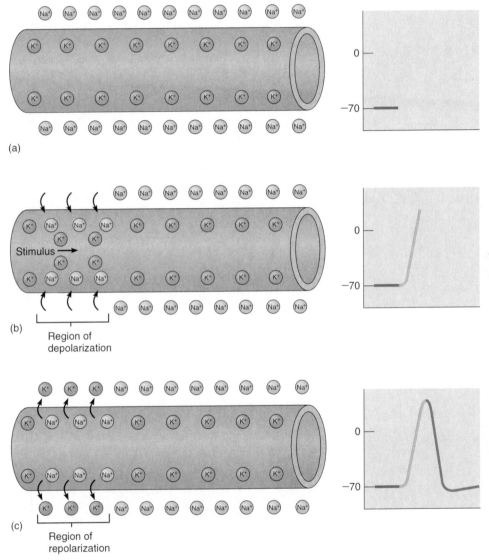

(a)

(b)

Region of
depolarization

(c)

Region of
repolarization

The Action Potential
Figures 10.12

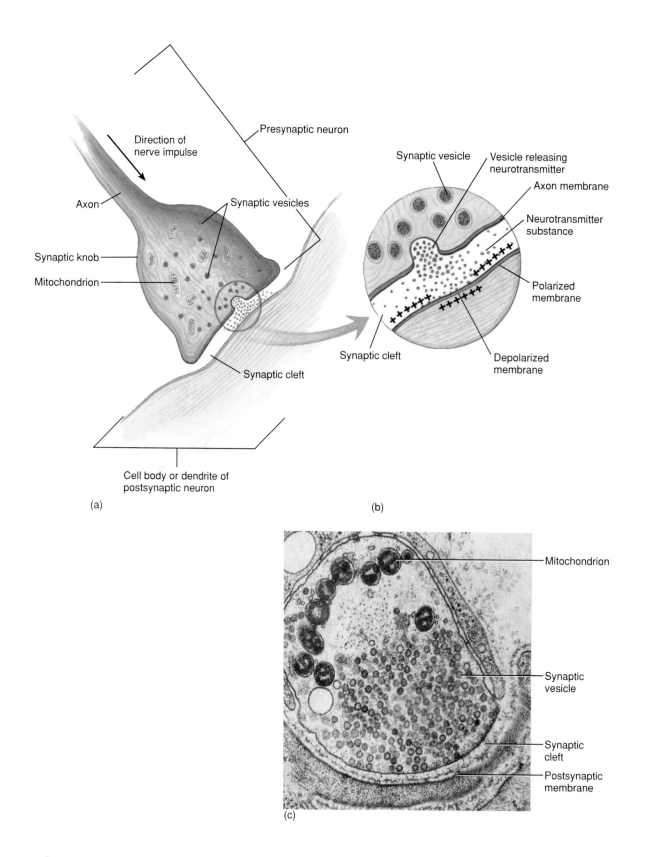

Direction of nerve impulse (a)

- Presynaptic neuron
- Axon
- Synaptic vesicles
- Synaptic knob
- Mitochondrion
- Synaptic cleft
- Cell body or dendrite of postsynaptic neuron

(b)

- Synaptic vesicle
- Vesicle releasing neurotransmitter
- Axon membrane
- Neurotransmitter substance
- Polarized membrane
- Depolarized membrane
- Synaptic cleft

(c)

- Mitochondrion
- Synaptic vesicle
- Synaptic cleft
- Postsynaptic membrane

Synaptic Knob and Synaptic Cleft
Figure 10.17

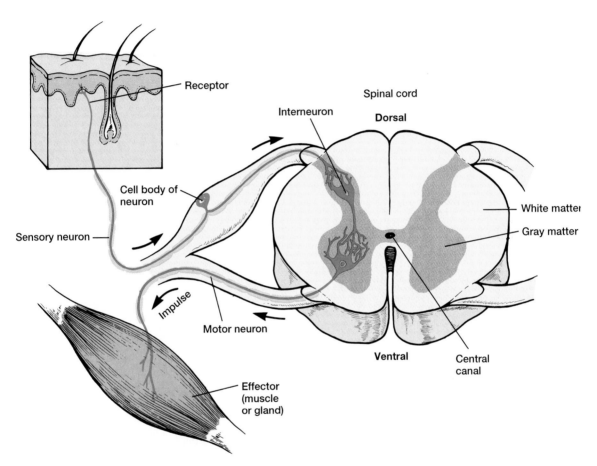

Receptor

Spinal cord

Dorsal

Interneuron

Cell body of neuron

Sensory neuron

White matter

Gray matter

Impulse

Motor neuron

Ventral

Central canal

Effector (muscle or gland)

Reflex Arc
Figure 10.21

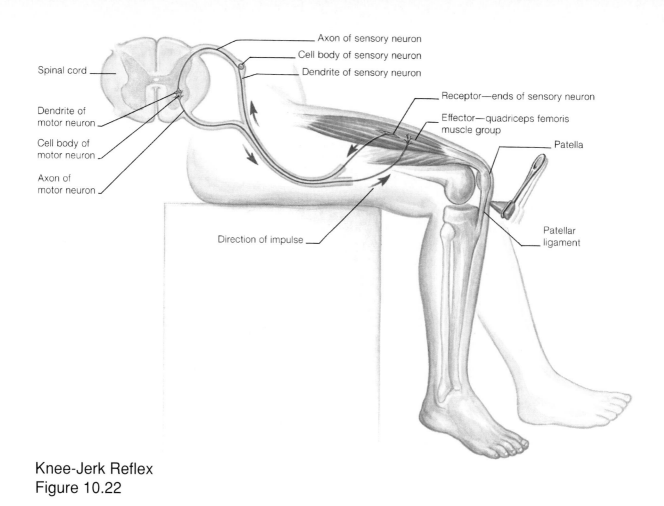

Knee-Jerk Reflex
Figure 10.22

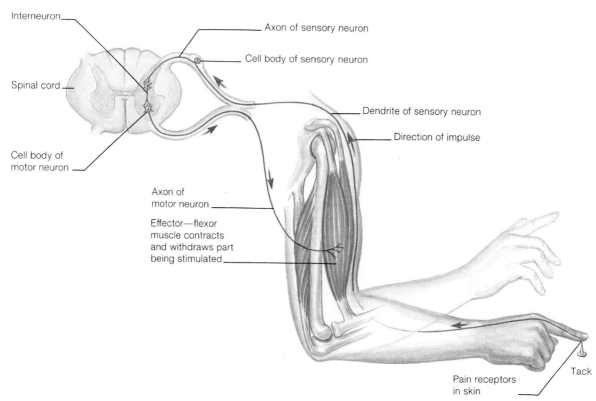

Withdrawal Reflex
Figure 10.23

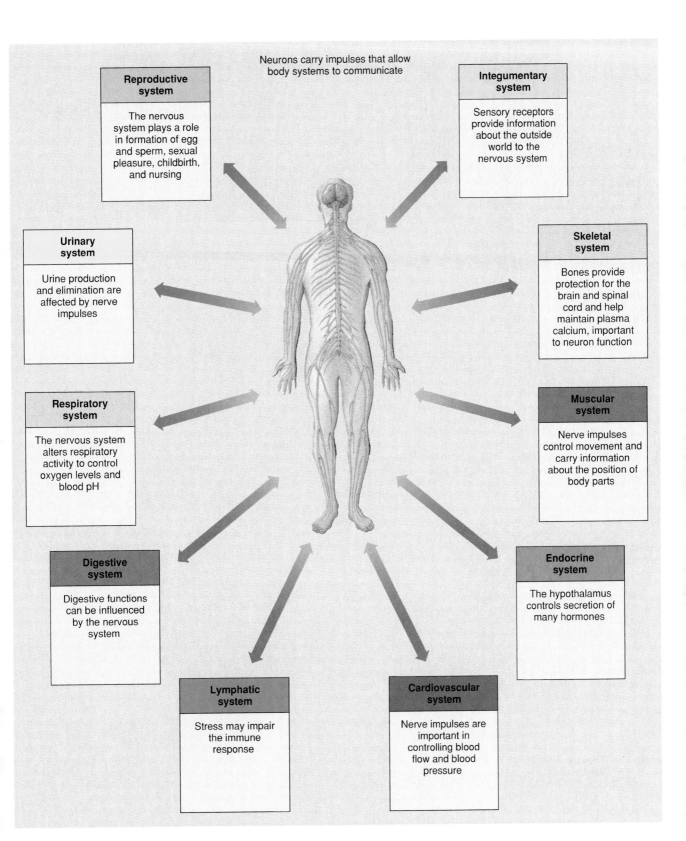

Neurons carry impulses that allow body systems to communicate

Reproductive system

The nervous system plays a role in formation of egg and sperm, sexual pleasure, childbirth, and nursing

Integumentary system

Sensory receptors provide information about the outside world to the nervous system

Urinary system

Urine production and elimination are affected by nerve impulses

Skeletal system

Bones provide protection for the brain and spinal cord and help maintain plasma calcium, important to neuron function

Respiratory system

The nervous system alters respiratory activity to control oxygen levels and blood pH

Muscular system

Nerve impulses control movement and carry information about the position of body parts

Digestive system

Digestive functions can be influenced by the nervous system

Endocrine system

The hypothalamus controls secretion of many hormones

Lymphatic system

Stress may impair the immune response

Cardiovascular system

Nerve impulses are important in controlling blood flow and blood pressure

Nervous System
InnerConnections: Chapter 10

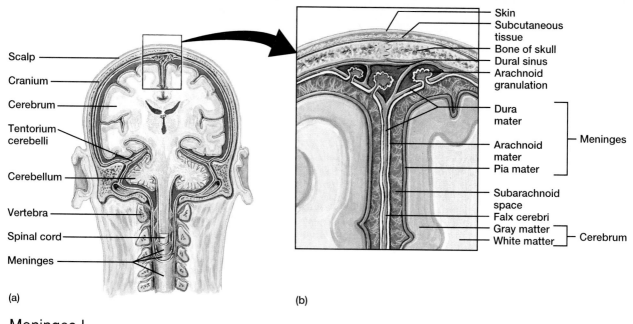

Scalp

Cranium

Cerebrum

Tentorium cerebelli

Cerebellum

Vertebra

Spinal cord

Meninges

(a)

Skin
Subcutaneous tissue
Bone of skull
Dural sinus
Arachnoid granulation

Dura mater

Arachnoid mater
Pia mater

Meninges

Subarachnoid space
Falx cerebri
Gray matter
White matter

Cerebrum

(b)

Meninges I
Figure 11.1

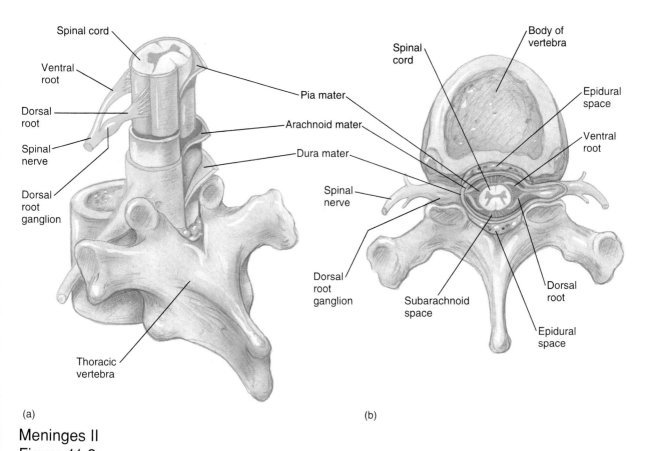

Spinal cord

Ventral root

Dorsal root

Spinal nerve

Dorsal root ganglion

Thoracic vertebra

(a)

Pia mater

Arachnoid mater

Dura mater

Spinal nerve

Dorsal root ganglion

Subarachnoid space

Spinal cord

Body of vertebra

Epidural space

Ventral root

Dorsal root

Epidural space

(b)

Meninges II
Figure 11.2

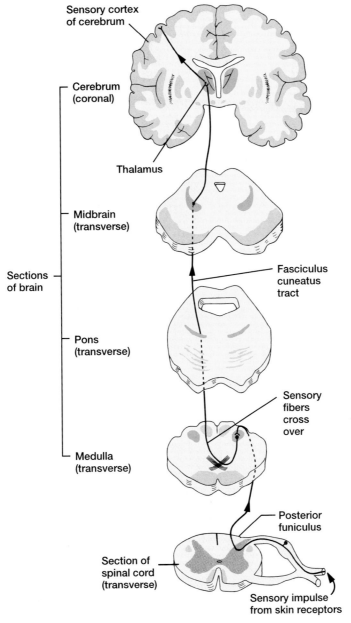

Sensory cortex
of cerebrum

Cerebrum
(coronal)

Thalamus

Midbrain
(transverse)

Sections
of brain

Fasciculus
cuneatus
tract

Pons
(transverse)

Sensory
fibers
cross
over

Medulla
(transverse)

Posterior
funiculus

Section of
spinal cord
(transverse)

Sensory impulse
from skin receptors

Ascending Tracts
Figure 11.6

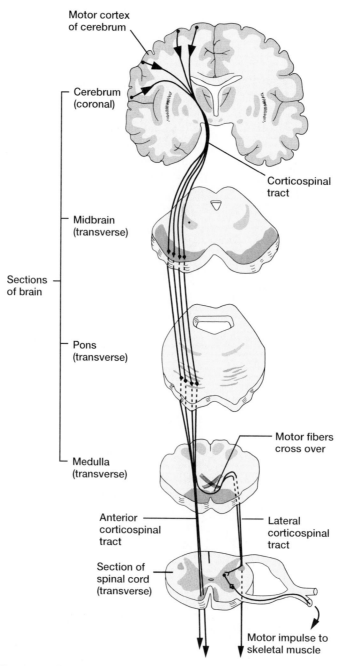

Motor cortex
of cerebrum

Cerebrum
(coronal)

Corticospinal
tract

Midbrain
(transverse)

Sections
of brain

Pons
(transverse)

Motor fibers
cross over

Medulla
(transverse)

Anterior
corticospinal
tract

Lateral
corticospinal
tract

Section of
spinal cord
(transverse)

Motor impulse to
skeletal muscle

Descending Tracts
Figure 11.7

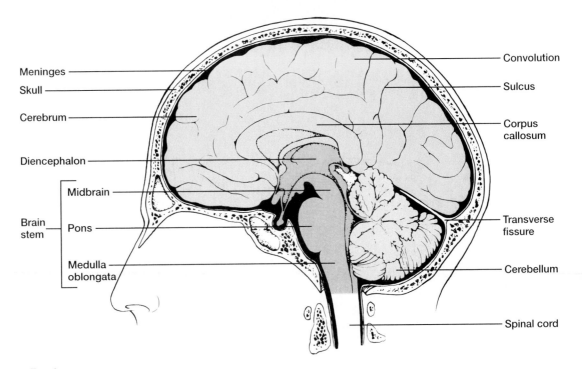

The Brain
Figure 11.9

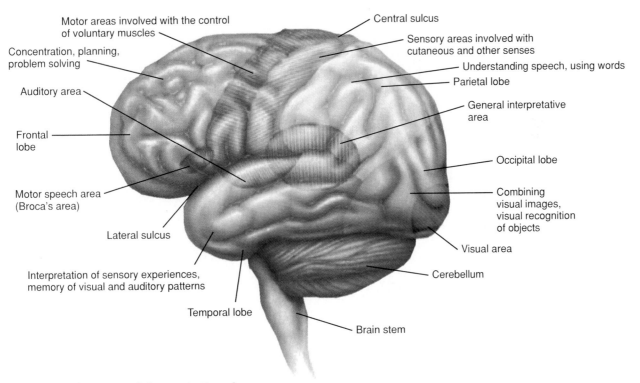

Sensory, Motor, and Association Areas
Figure 11.11

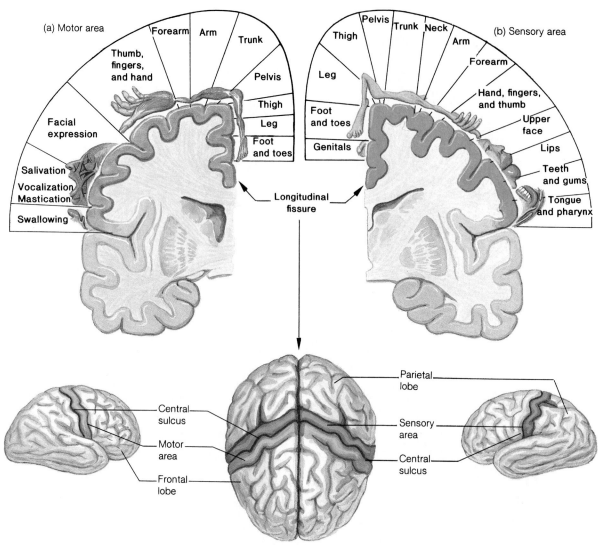

(a) Motor area

Forearm | Arm | Trunk

Thumb, fingers, and hand

Pelvis

Thigh

Leg

Foot and toes

Facial expression

Salivation

Vocalization

Mastication

Swallowing

(b) Sensory area

Pelvis | Trunk | Neck

Thigh

Arm

Leg

Forearm

Foot and toes

Hand, fingers, and thumb

Genitals

Upper face

Lips

Teeth and gums

Tongue and pharynx

Longitudinal fissure

Parietal lobe

Central sulcus

Sensory area

Motor area

Central sulcus

Frontal lobe

Motor and Sensory Areas
Figure 11.12

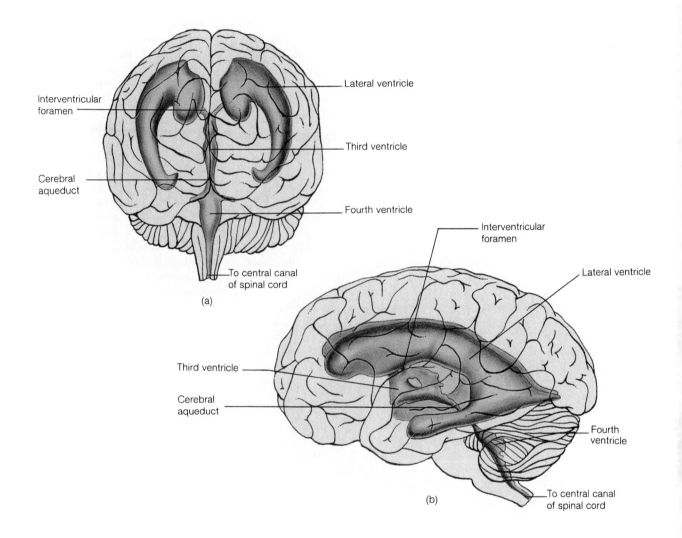

Interventricular foramen

Lateral ventricle

Third ventricle

Cerebral aqueduct

Fourth ventricle

To central canal of spinal cord

(a)

Interventricular foramen

Lateral ventricle

Third ventricle

Cerebral aqueduct

Fourth ventricle

To central canal of spinal cord

(b)

Ventricles of the Brain
Figure 11.14

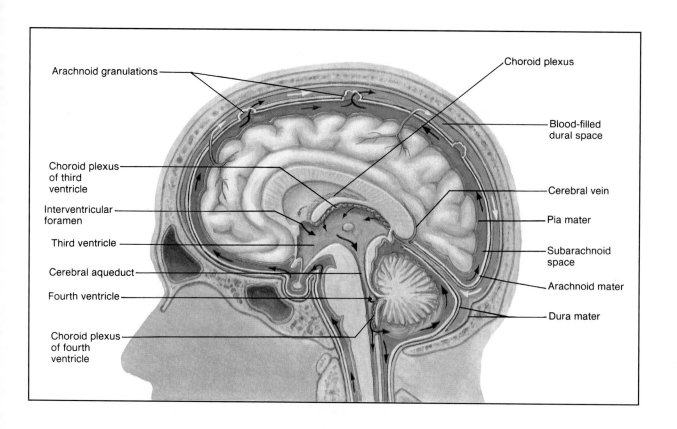

Arachnoid granulations

Choroid plexus

Choroid plexus
of third
ventricle

Interventricular
foramen

Third ventricle

Cerebral aqueduct

Fourth ventricle

Choroid plexus
of fourth
ventricle

Blood-filled
dural space

Cerebral vein

Pia mater

Subarachnoid
space

Arachnoid mater

Dura mater

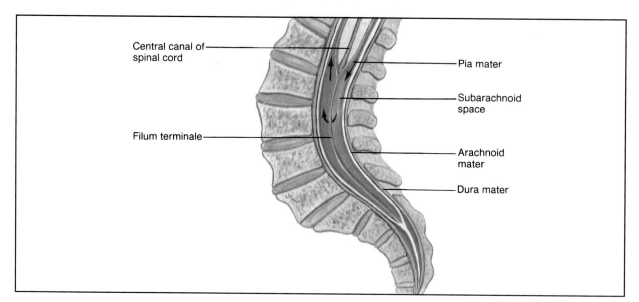

Central canal of
spinal cord

Filum terminale

Pia mater

Subarachnoid
space

Arachnoid
mater

Dura mater

Cerebrospinal Fluid Circulation
Figure 11.15

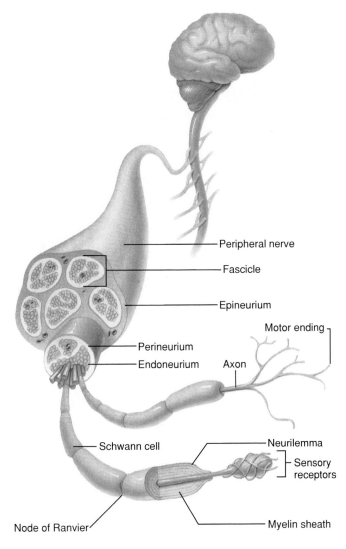

Peripheral nerve

Fascicle

Epineurium

Motor ending

Perineurium

Axon

Endoneurium

Schwann cell

Neurilemma

Sensory receptors

Node of Ranvier

Myelin sheath

A Mixed Nerve
Figure 11.20

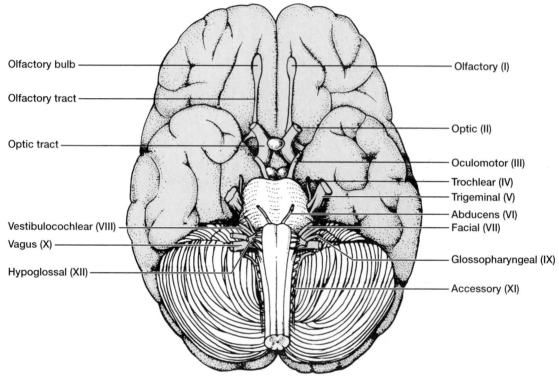

Olfactory bulb ——————————— Olfactory (I)

Olfactory tract ——————

Optic tract —————— Optic (II)

Oculomotor (III)
Trochlear (IV)
Trigeminal (V)

Vestibulocochlear (VIII) —————— Abducens (VI)
Facial (VII)

Vagus (X) —————————

Hypoglossal (XII) —————— Glossopharyngeal (IX)

Accessory (XI)

Cranial Nerves
Figure 11.22

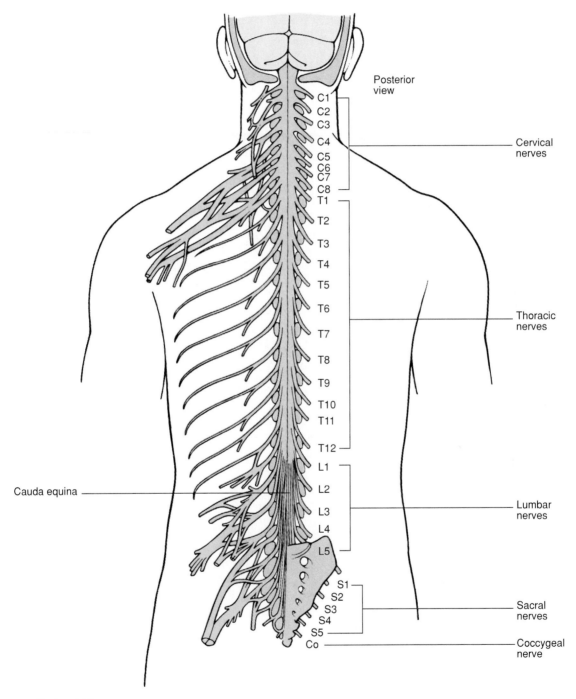

Posterior
view

C1
C2
C3
C4 — Cervical nerves
C5
C6
C7
C8

T1
T2
T3
T4
T5
T6 — Thoracic nerves
T7
T8
T9
T10
T11
T12

L1
L2
L3 — Lumbar nerves
L4
L5

Cauda equina

S1
S2
S3 — Sacral nerves
S4
S5
Co — Coccygeal nerve

Spinal Nerves
Figure 11.26

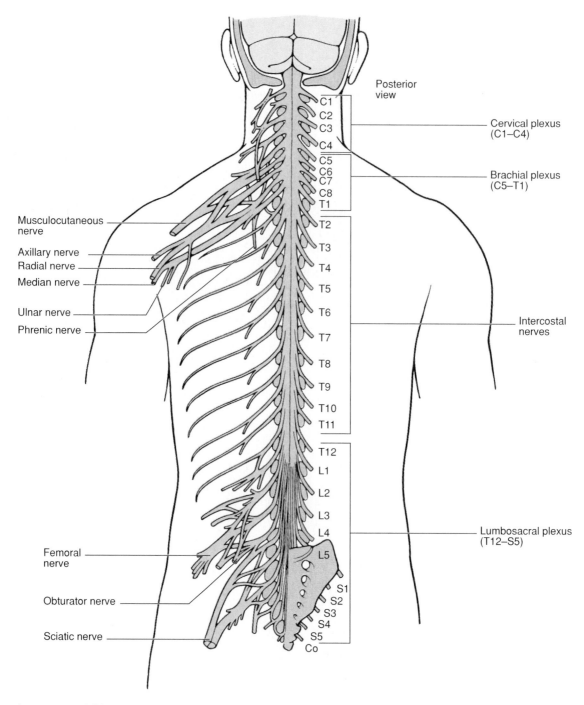

C1
C2
C3
C4
C5
C6
C7
C8
T1
T2
T3
T4
T5
T6
T7
T8
T9
T10
T11
T12
L1
L2
L3
L4
L5
S1
S2
S3
S4
S5
Co

Posterior view

Cervical plexus (C1–C4)

Brachial plexus (C5–T1)

Intercostal nerves

Lumbosacral plexus (T12–S5)

Musculocutaneous nerve

Axillary nerve

Radial nerve

Median nerve

Ulnar nerve

Phrenic nerve

Femoral nerve

Obturator nerve

Sciatic nerve

Intercostal Nerves
Figure 11.29

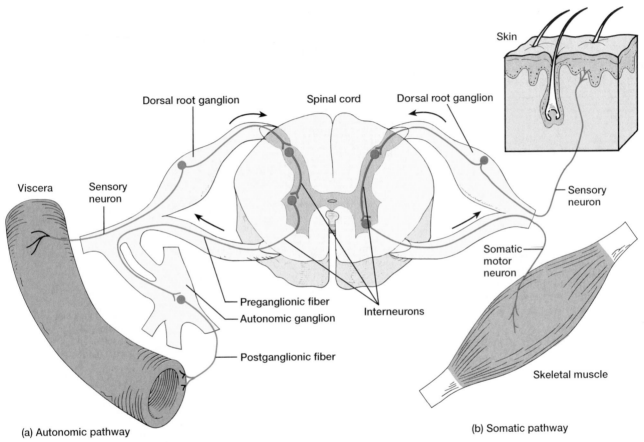

Dorsal root ganglion · Spinal cord · Dorsal root ganglion · Skin · Viscera · Sensory neuron · Sensory neuron · Somatic motor neuron · Preganglionic fiber · Autonomic ganglion · Interneurons · Postganglionic fiber · Skeletal muscle

(a) Autonomic pathway

(b) Somatic pathway

Somatic and Autonomic Nerve Pathways
Figure 11.32

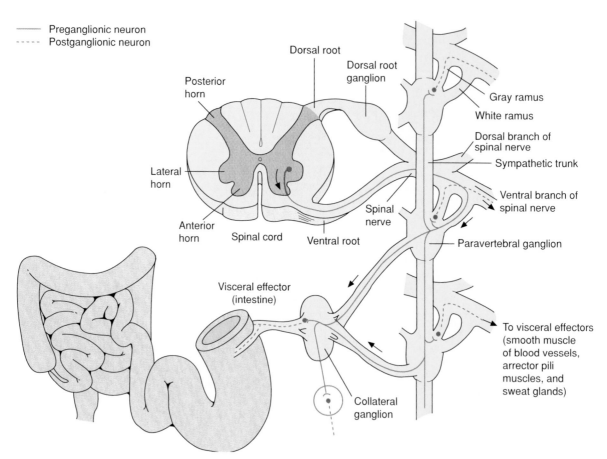

Legend:
——— Preganglionic neuron
- - - Postganglionic neuron

Posterior horn
Dorsal root
Dorsal root ganglion
Gray ramus
White ramus
Dorsal branch of spinal nerve
Sympathetic trunk
Lateral horn
Ventral branch of spinal nerve
Anterior horn
Spinal cord
Ventral root
Spinal nerve
Paravertebral ganglion
Visceral effector (intestine)
To visceral effectors (smooth muscle of blood vessels, arrector pili muscles, and sweat glands)
Collateral ganglion

Sympathetic Fibers in Spinal Nerves
Figure 11.34

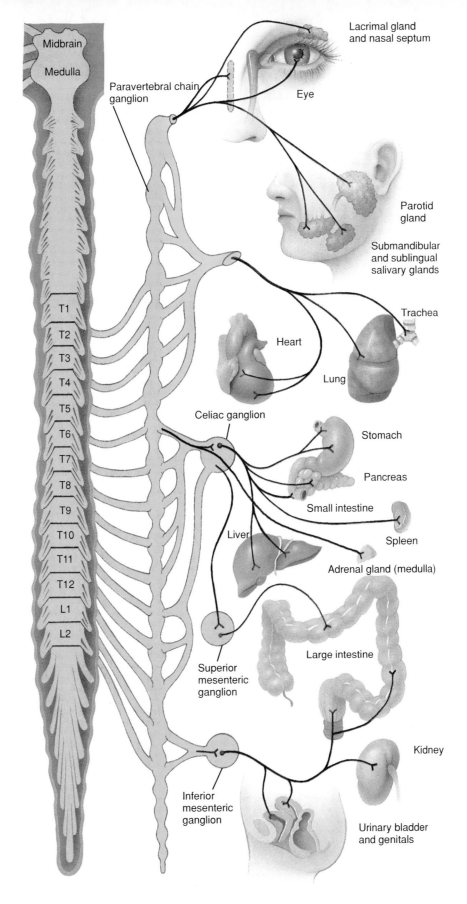

Sympathetic Nervous System
Figure 11.35

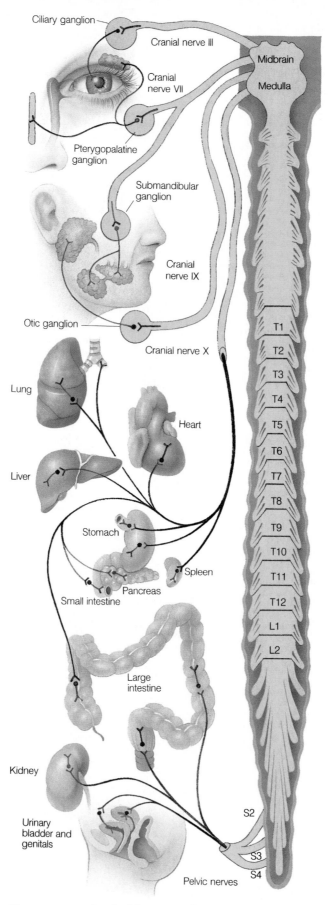

Parasympathetic Nervous System
Figure 11.36

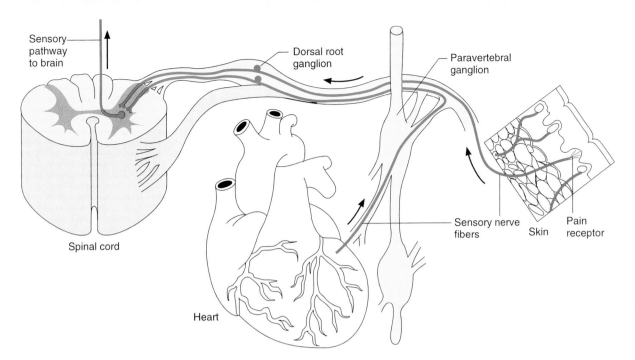

Referred Pain Pathway
Figure 12.4

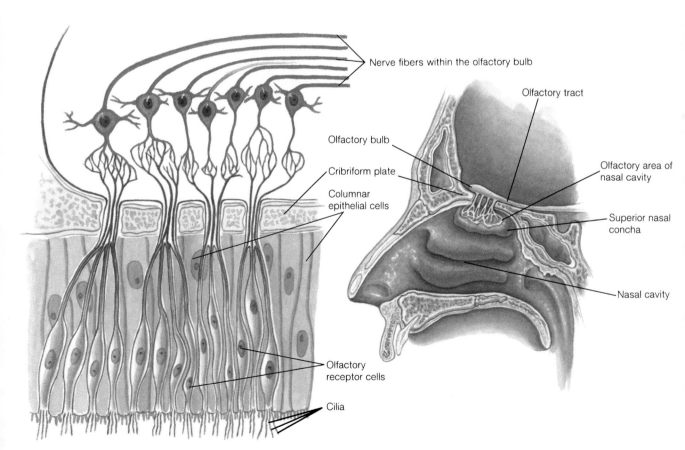

Olfactory Receptors
Figure 12.6

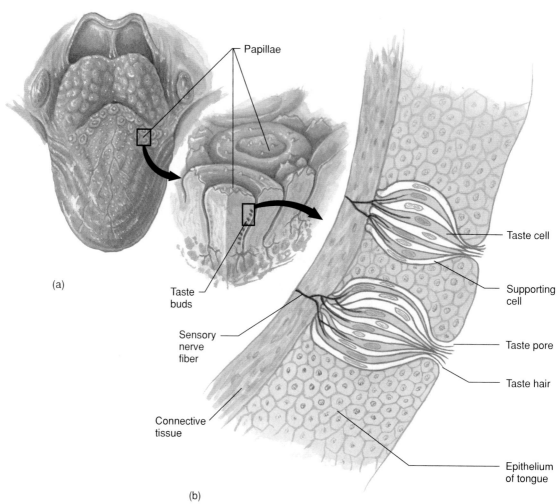

(a)

(b)

Papillae

Taste buds

Sensory nerve fiber

Connective tissue

Taste cell

Supporting cell

Taste pore

Taste hair

Epithelium of tongue

Taste Receptors
Figure 12.8

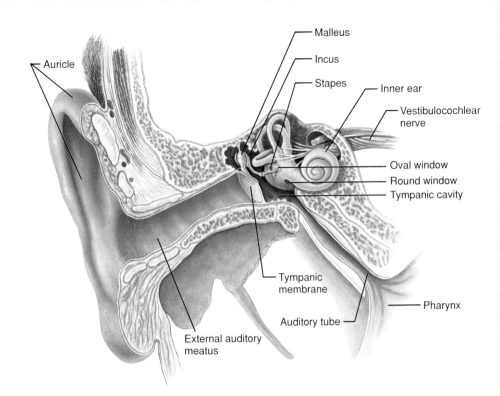

Structure of the Ear
Figure 12.11

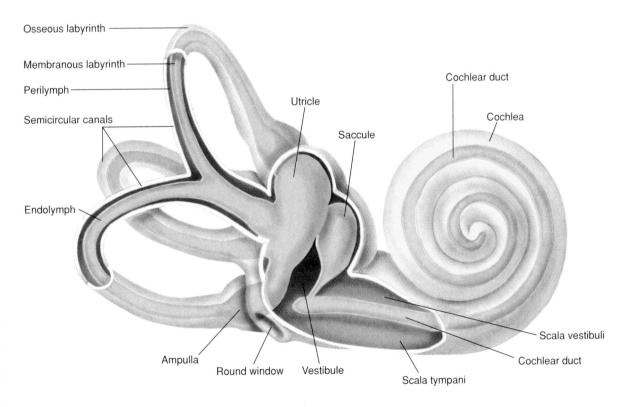

Inner Ear Structure
Figure 12.13

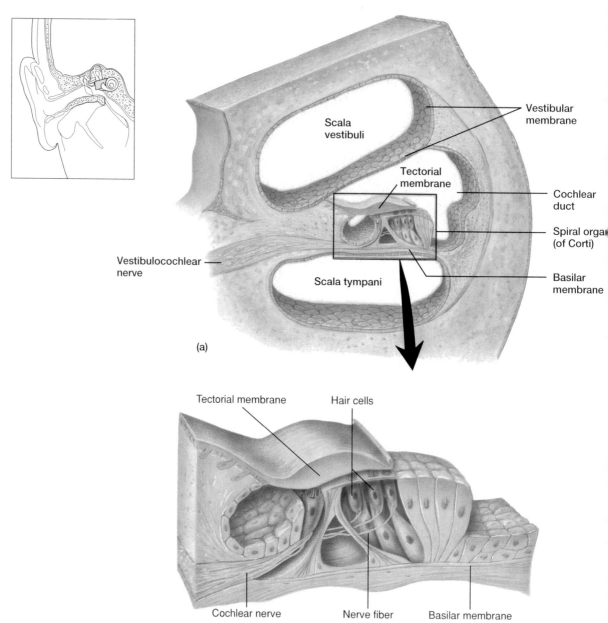

(a)

Scala
vestibuli

Vestibular
membrane

Tectorial
membrane

Cochlear
duct

Spiral organ
(of Corti)

Basilar
membrane

Vestibulocochlear
nerve

Scala tympani

Tectorial membrane

Hair cells

Cochlear nerve

Nerve fiber

Basilar membrane

Organ of Corti
Figure 12.15

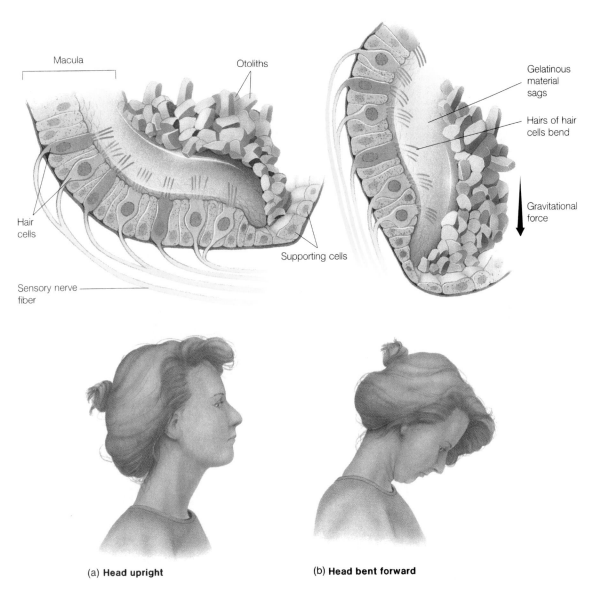

Macula

Otoliths

Gelatinous
material
sags

Hairs of hair
cells bend

Hair
cells

Supporting cells

Gravitational
force

Sensory nerve
fiber

(a) **Head upright**

(b) **Head bent forward**

Static Equilibrium
Figure 12.20

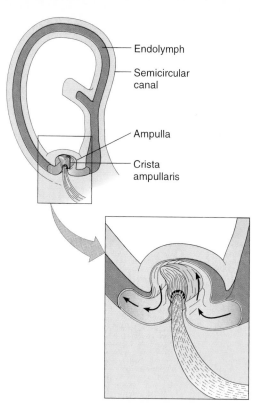

Endolymph

Semicircular canal

Ampulla

Crista ampullaris

Dynamic Equilibrium
Figure 12.23

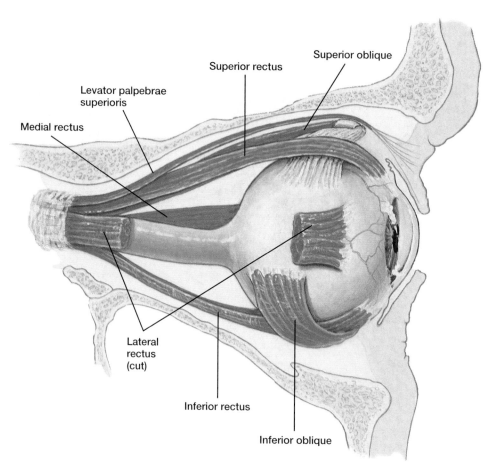

Superior rectus

Superior oblique

Levator palpebrae superioris

Medial rectus

Lateral rectus (cut)

Inferior rectus

Inferior oblique

Extrinsic Muscles of the Eye
Figure 12.26

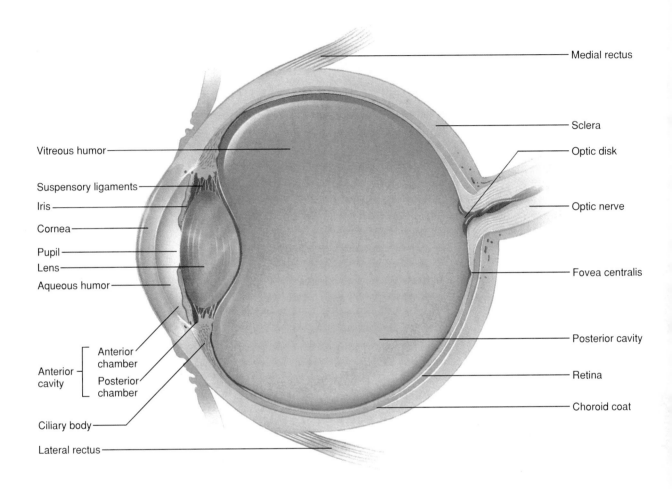

Medial rectus

Sclera

Optic disk

Vitreous humor

Optic nerve

Suspensory ligaments

Iris

Cornea

Pupil

Lens

Aqueous humor

Fovea centralis

Posterior cavity

Anterior
chamber

Anterior
cavity

Posterior
chamber

Retina

Choroid coat

Ciliary body

Lateral rectus

Eye, Transverse Section
Figure 12.27

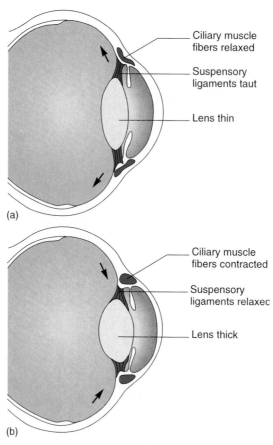

(a)

Ciliary muscle
fibers relaxed

Suspensory
ligaments taut

Lens thin

(b)

Ciliary muscle
fibers contracted

Suspensory
ligaments relaxed

Lens thick

Lens and Ciliary Body
Figures 12.31

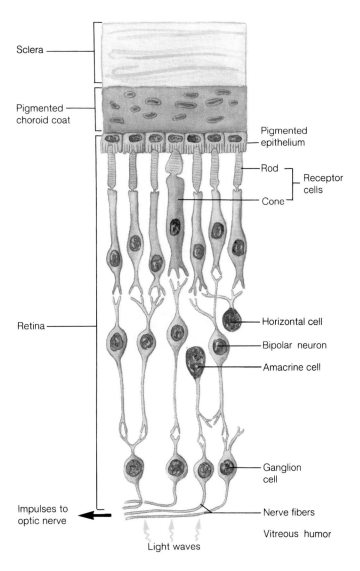

Sclera

Pigmented choroid coat

Pigmented epithelium

Rod ⎱ Receptor
Cone ⎰ cells

Retina

Horizontal cell

Bipolar neuron

Amacrine cell

Ganglion cell

Impulses to optic nerve

Nerve fibers

Vitreous humor

Light waves

Retina
Figure 12.34

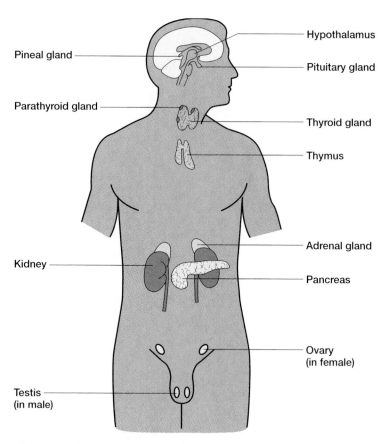

Pineal gland

Parathyroid gland

Kidney

Testis
(in male)

Hypothalamus

Pituitary gland

Thyroid gland

Thymus

Adrenal gland

Pancreas

Ovary
(in female)

Major Endocrine Glands
Figure 13.2

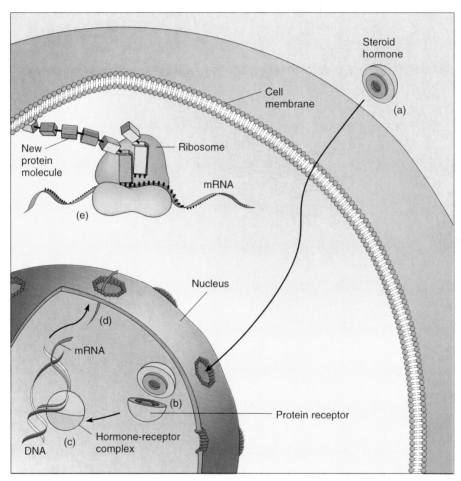

Steroid Hormone Action
Figure 13.4

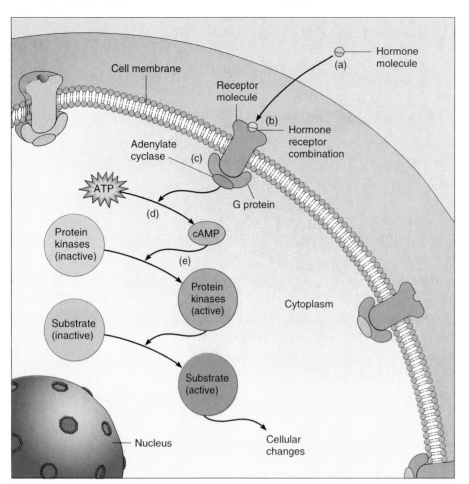

Nonsteroid Hormone Action
Figure 13.6

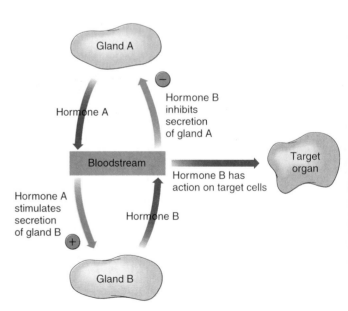

Negative Feedback System
Figure 13.7

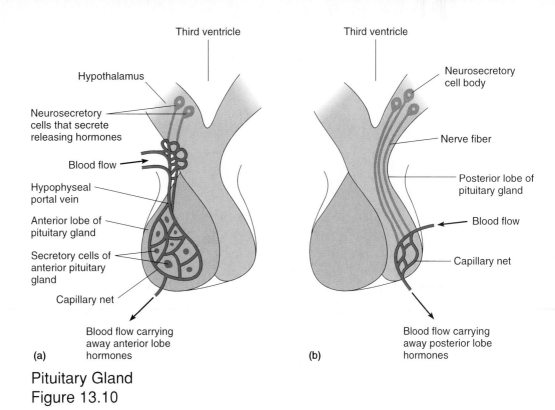

Pituitary Gland
Figure 13.10

Third ventricle

Hypothalamus

Neurosecretory cells that secrete releasing hormones

Blood flow

Hypophyseal portal vein

Anterior lobe of pituitary gland

Secretory cells of anterior pituitary gland

Capillary net

(a)

Blood flow carrying away anterior lobe hormones

Third ventricle

Neurosecretory cell body

Nerve fiber

Posterior lobe of pituitary gland

Blood flow

Capillary net

(b)

Blood flow carrying away posterior lobe hormones

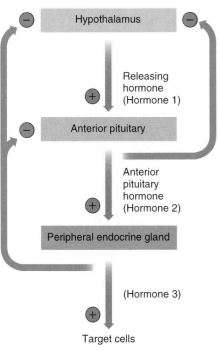

Hypothalamus

Releasing hormone (Hormone 1)

Anterior pituitary

Anterior pituitary hormone (Hormone 2)

Peripheral endocrine gland

(Hormone 3)

Target cells

Hypothalmic Control of Hormone Secretion
Figure 13.11

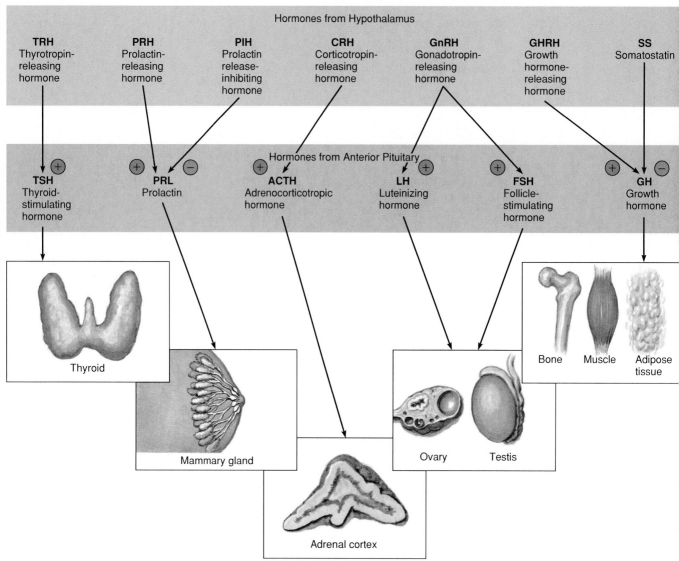

Hormones from the Hypothalamus and Anterior Pituitary
Figure 13.13

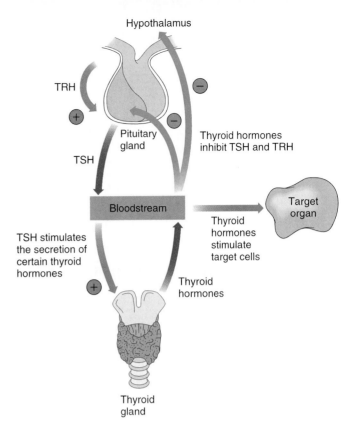

Hypothalamus

TRH

Pituitary
gland

Thyroid hormones
inhibit TSH and TRH

TSH

Bloodstream

Target
organ

TSH stimulates
the secretion of
certain thyroid
hormones

Thyroid
hormones
stimulate
target cells

Thyroid
hormones

Thyroid
gland

Control of Thyroid Hormone Secretion
Figure 13.14

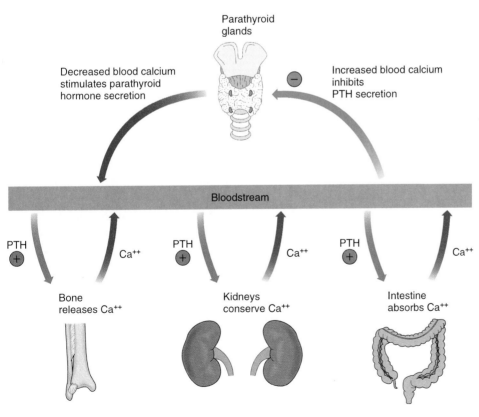

Parathyroid
glands

Decreased blood calcium
stimulates parathyroid
hormone secretion

Increased blood calcium
inhibits
PTH secretion

Bloodstream

PTH

Ca++

PTH

Ca++

PTH

Ca++

Bone
releases Ca++

Kidneys
conserve Ca++

Intestine
absorbs Ca++

Parathyroid Hormone
Figure 13.25

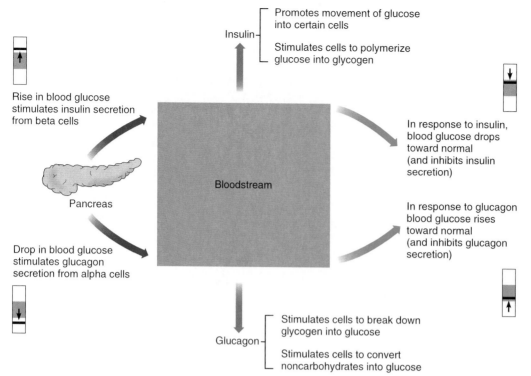

Insulin ⎯ Promotes movement of glucose into certain cells

Stimulates cells to polymerize glucose into glycogen

Rise in blood glucose stimulates insulin secretion from beta cells

Bloodstream

Pancreas

In response to insulin, blood glucose drops toward normal (and inhibits insulin secretion)

In response to glucagon blood glucose rises toward normal (and inhibits glucagon secretion)

Drop in blood glucose stimulates glucagon secretion from alpha cells

Glucagon ⎯ Stimulates cells to break down glycogen into glucose

Stimulates cells to convert noncarbohydrates into glucose

Control of Blood Glucose
Figure 13.34

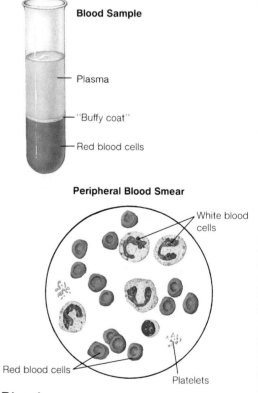

Blood Sample

Plasma

"Buffy coat"

Red blood cells

Peripheral Blood Smear

White blood cells

Red blood cells

Platelets

Blood
Figure 14.1

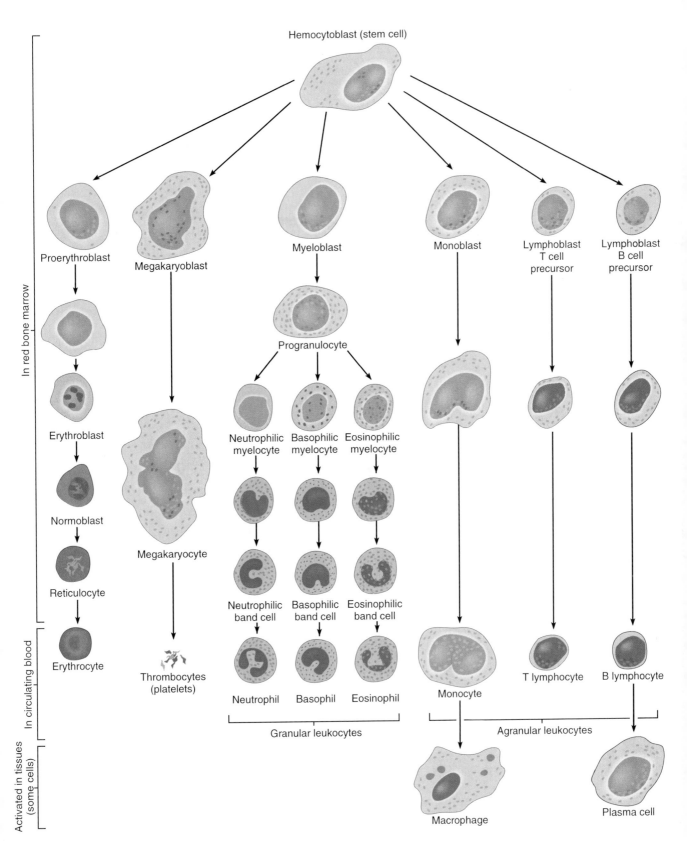

Hemocytoblast (stem cell)

In red bone marrow

Proerythroblast

Erythroblast

Normoblast

Reticulocyte

In circulating blood

Erythrocyte

Megakaryoblast

Megakaryocyte

Thrombocytes
(platelets)

Myeloblast

Progranulocyte

Neutrophilic
myelocyte

Basophilic
myelocyte

Eosinophilic
myelocyte

Neutrophilic
band cell

Basophilic
band cell

Eosinophilic
band cell

Neutrophil

Basophil

Eosinophil

Granular leukocytes

Monoblast

Monocyte

Activated in tissues
(some cells)

Macrophage

Lymphoblast
T cell
precursor

Lymphoblast
B cell
precursor

T lymphocyte

B lymphocyte

Agranular leukocytes

Plasma cell

Blood Cell Development
Figure 14.3a

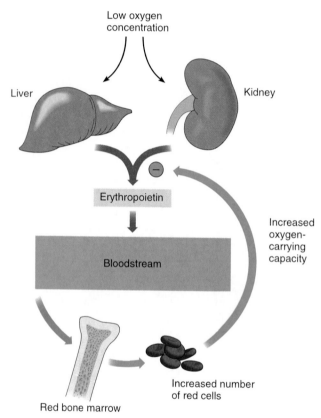

Low oxygen
concentration

Liver

Kidney

Erythropoietin

Bloodstream

Increased
oxygen-
carrying
capacity

Red bone marrow

Increased number
of red cells

Low Oxygen Pressure
Figure 14.6

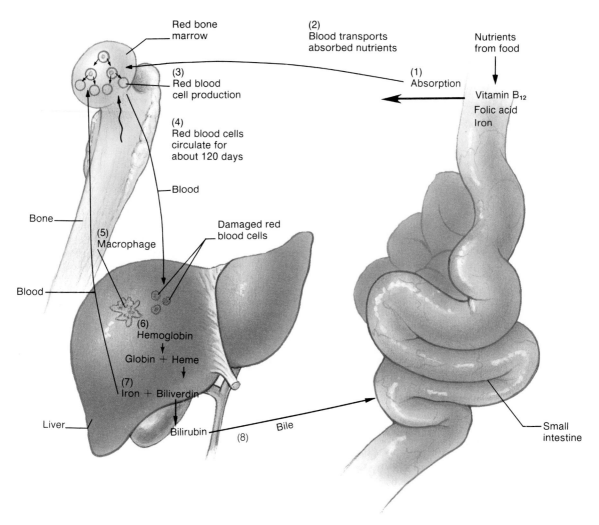

Red Blood Cells
Figure 14.7

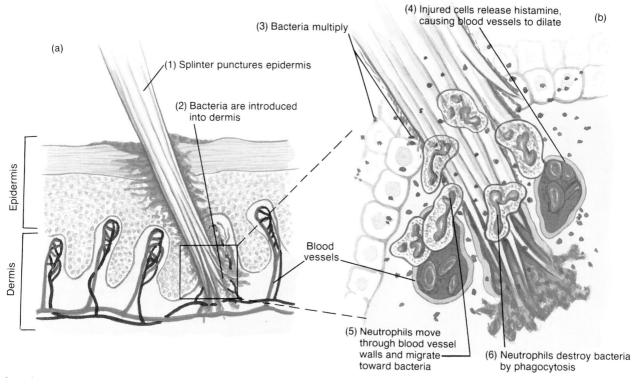

(a)

(1) Splinter punctures epidermis

(2) Bacteria are introduced into dermis

(3) Bacteria multiply

(4) Injured cells release histamine, causing blood vessels to dilate

(b)

Epidermis

Dermis

Blood vessels

(5) Neutrophils move through blood vessel walls and migrate toward bacteria

(6) Neutrophils destroy bacteria by phagocytosis

Leukocytes
Figure 14.15

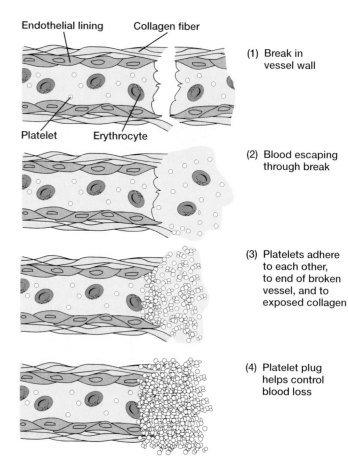

Endothelial lining   Collagen fiber

Platelet   Erythrocyte

(1) Break in vessel wall

(2) Blood escaping through break

(3) Platelets adhere to each other, to end of broken vessel, and to exposed collagen

(4) Platelet plug helps control blood loss

Platelet Plug Formation
Figure 14.17

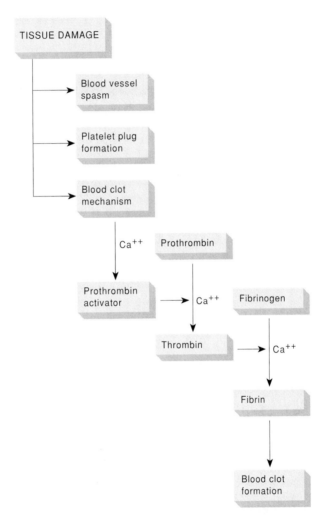

Hemostasis
Figure 14.18

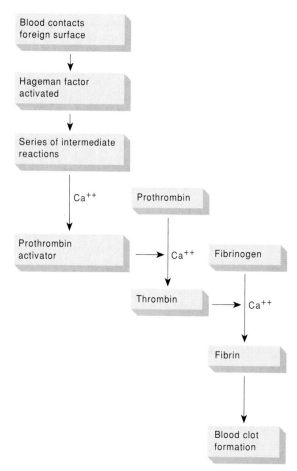

Intrinsic Blood Clotting Mechanism
Figure 14.20

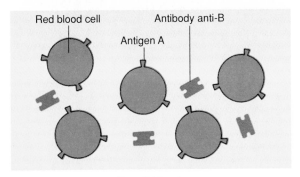

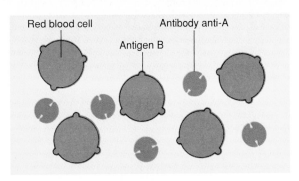

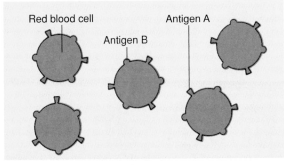

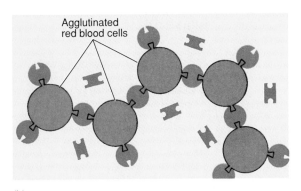

ABO Blood Types
Figures 14.22

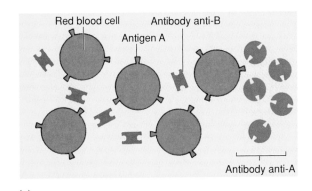

(a)

(b)

ABO Blood Reactions
Figure 14.23

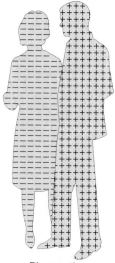

Rh-negative
woman and
Rh-positive man
conceive a child.

Rh-negative
woman with
Rh-positive
fetus.

Cells from
Rh-positive
fetus enter
mother's
bloodstream.

Woman
becomes
sensitized—
antibodies (⊕)
form to fight
Rh-positive
blood cells.

In the next
Rh-positive
pregnancy,
antibodies
attack fetal
blood cells.

Rh⁺ Factor
Figure 14.24

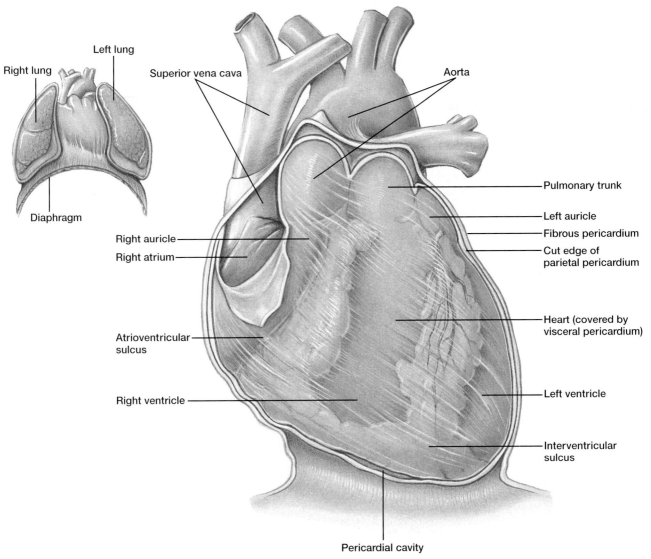

Right lung   Left lung

Superior vena cava   Aorta

Diaphragm

Pulmonary trunk

Right auricle

Right atrium

Left auricle

Fibrous pericardium

Cut edge of
parietal pericardium

Atrioventricular
sulcus

Heart (covered by
visceral pericardium)

Right ventricle

Left ventricle

Interventricular
sulcus

Pericardial cavity

The Heart
Figure 15.4

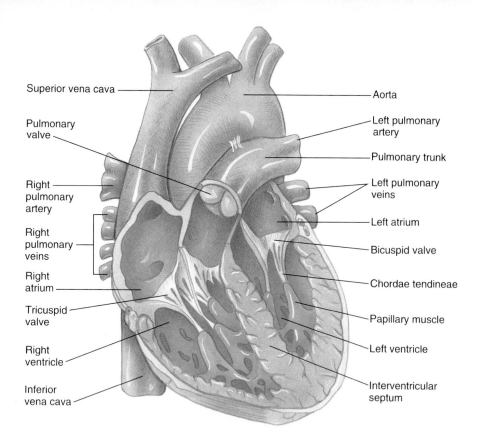

Superior vena cava

Pulmonary valve

Right pulmonary artery

Right pulmonary veins

Right atrium

Tricuspid valve

Right ventricle

Inferior vena cava

Aorta

Left pulmonary artery

Pulmonary trunk

Left pulmonary veins

Left atrium

Bicuspid valve

Chordae tendineae

Papillary muscle

Left ventricle

Interventricular septum

Heart, Frontal Section I
Figure 15.6a

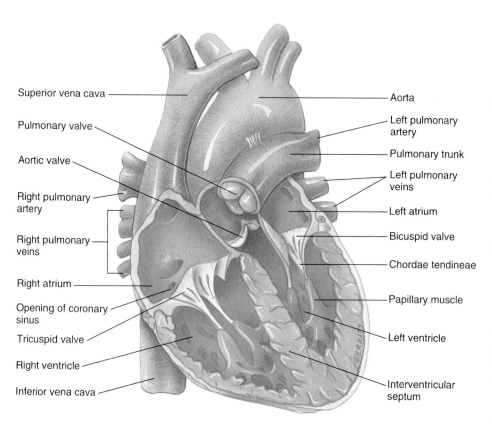

Superior vena cava

Pulmonary valve

Aortic valve

Right pulmonary artery

Right pulmonary veins

Right atrium

Opening of coronary sinus

Tricuspid valve

Right ventricle

Inferior vena cava

Aorta

Left pulmonary artery

Pulmonary trunk

Left pulmonary veins

Left atrium

Bicuspid valve

Chordae tendineae

Papillary muscle

Left ventricle

Interventricular septum

Heart, Frontal Section II
Figure 15.6b

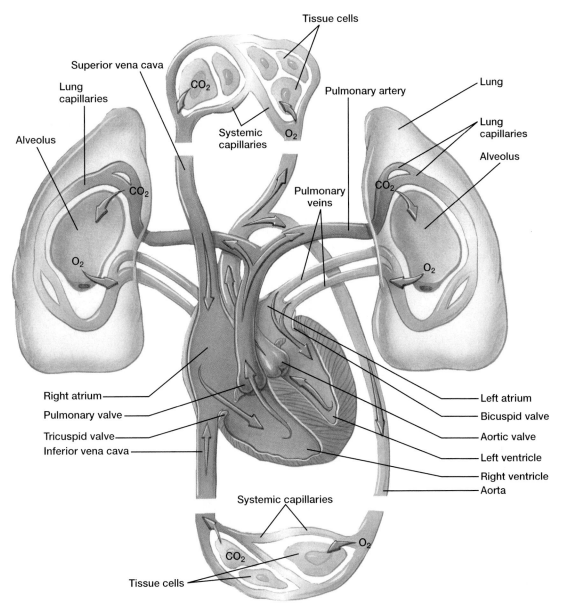

The Right Ventricle
Figure 15.10

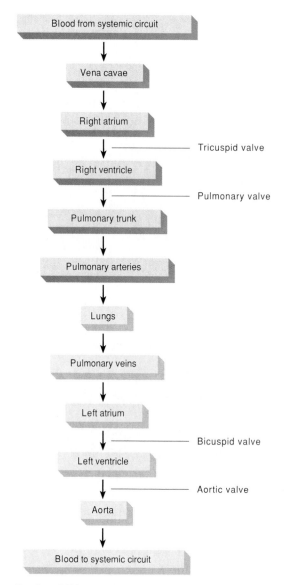

Path of Blood through the Heart
Figure 15.11

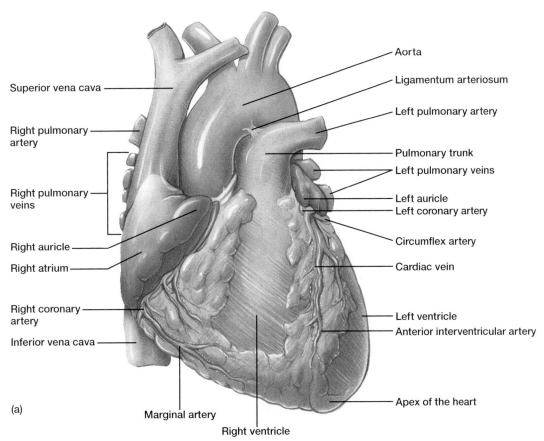

Superior vena cava

Right pulmonary artery

Right pulmonary veins

Right auricle

Right atrium

Right coronary artery

Inferior vena cava

(a)

Aorta

Ligamentum arteriosum

Left pulmonary artery

Pulmonary trunk

Left pulmonary veins

Left auricle

Left coronary artery

Circumflex artery

Cardiac vein

Left ventricle

Anterior interventricular artery

Apex of the heart

Marginal artery

Right ventricle

Heart and Coronary Vessels, Anterior
Figure 15.12a

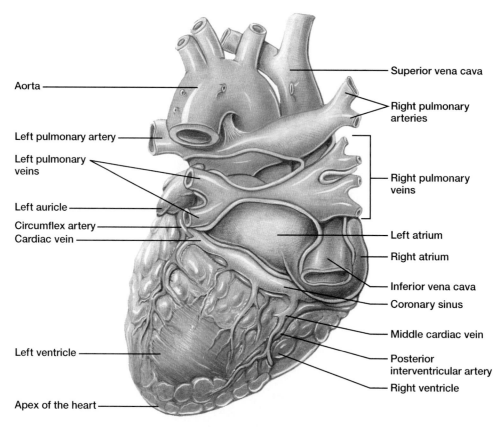

Aorta

Left pulmonary artery

Left pulmonary veins

Left auricle

Circumflex artery

Cardiac vein

Left ventricle

Apex of the heart

Superior vena cava

Right pulmonary arteries

Right pulmonary veins

Left atrium

Right atrium

Inferior vena cava

Coronary sinus

Middle cardiac vein

Posterior interventricular artery

Right ventricle

Heart and Coronary Vessels, Posterior
Figure 15.12b

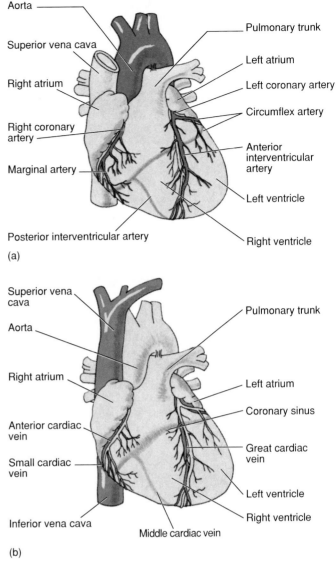

(a)

(b)

Coronary Arteries and Cardiac Veins
Figure 15.13

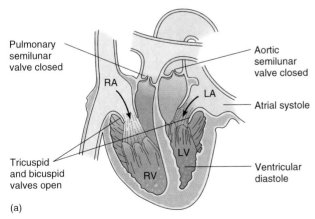

(a)

Atrial Systole and Diastole
Figure 15.16

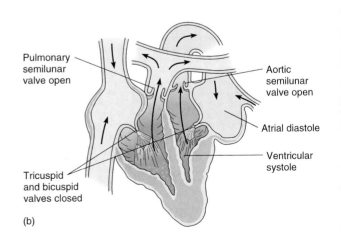

(b)

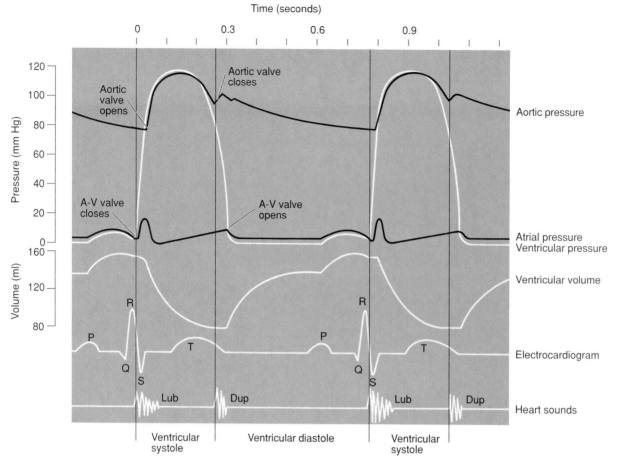

Cardiac Cycle
Figure 15.17

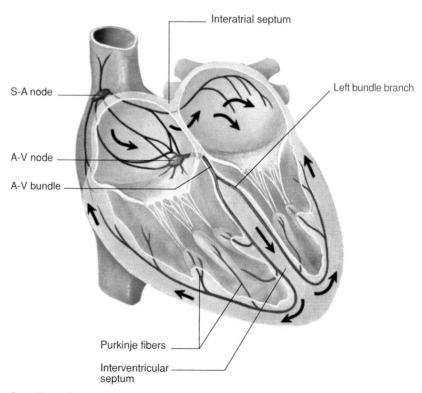

Cardiac Conduction System
Figure 15.19

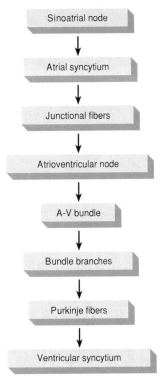

Cardiac Conduction System Components
Figure 15.20

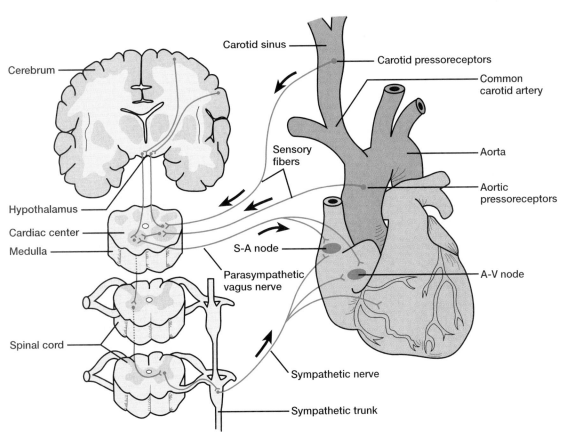

Heart Rate Control
Figure 15.22

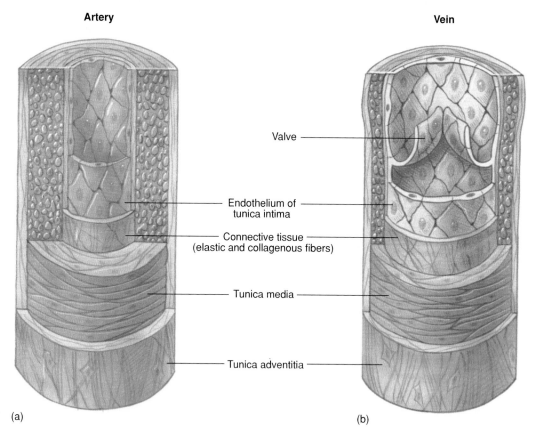

Artery

Vein

Valve

Endothelium of
tunica intima

Connective tissue
(elastic and collagenous fibers)

Tunica media

Tunica adventitia

(a)

(b)

Wall of an Artery and a Vein
Figure 15.23

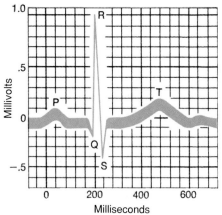

Depolarization and Repolarization
Box 15.2, Figure 15D

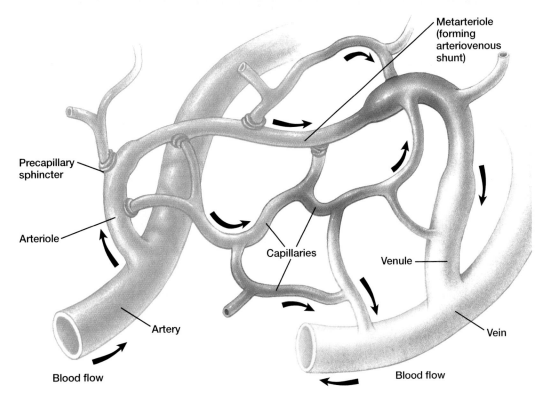

Metarterioles
Figure 15.26

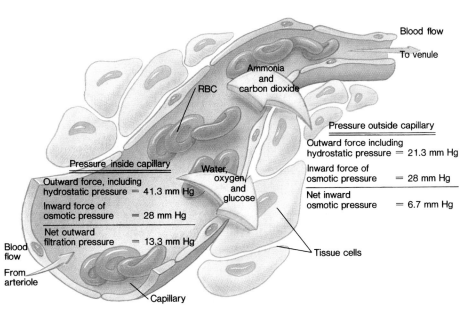

Capillary Exchanges
Figure 15.29

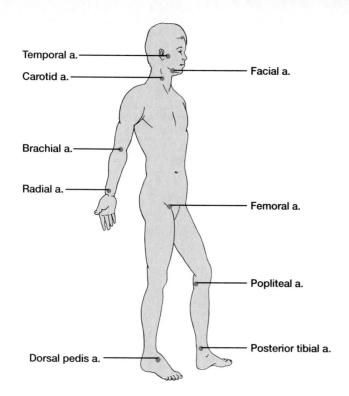

Temporal a.
Carotid a.
Facial a.
Brachial a.
Radial a.
Femoral a.
Popliteal a.
Posterior tibial a.
Dorsal pedis a.

Arterial Pulse Sites
Figure 15.33

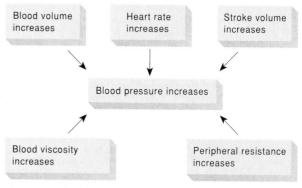

Blood volume increases

Heart rate increases

Stroke volume increases

Blood pressure increases

Blood viscosity increases

Peripheral resistance increases

Arterial Blood Pressure
Figure 15.34

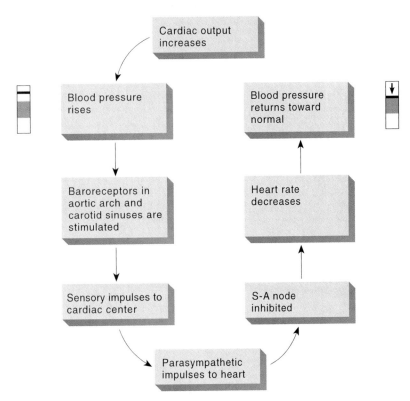

Inhibiting the S-A Node
Figure 15.37

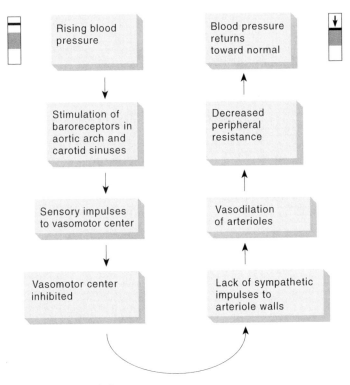

Dilating Arterioles
Figure 15.38

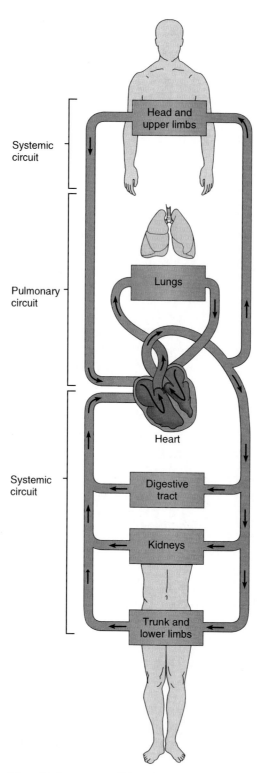

The Pulmonary Circuit
Figure 15.40

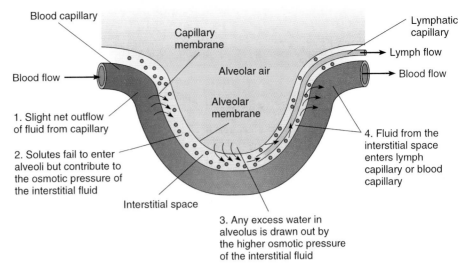

Cells of the Alveolar Wall
Figure 15.42

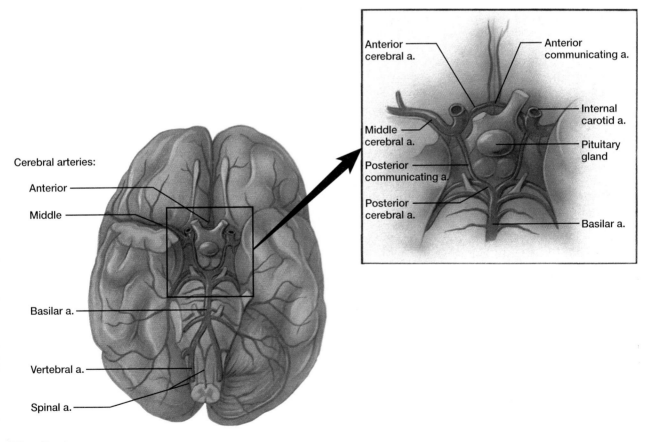

Cerebral arteries:

Anterior

Middle

Basilar a.

Vertebral a.

Spinal a.

Anterior cerebral a.

Anterior communicating a.

Middle cerebral a.

Internal carotid a.

Posterior communicating a.

Pituitary gland

Posterior cerebral a.

Basilar a.

The Brain
Figure 15.47

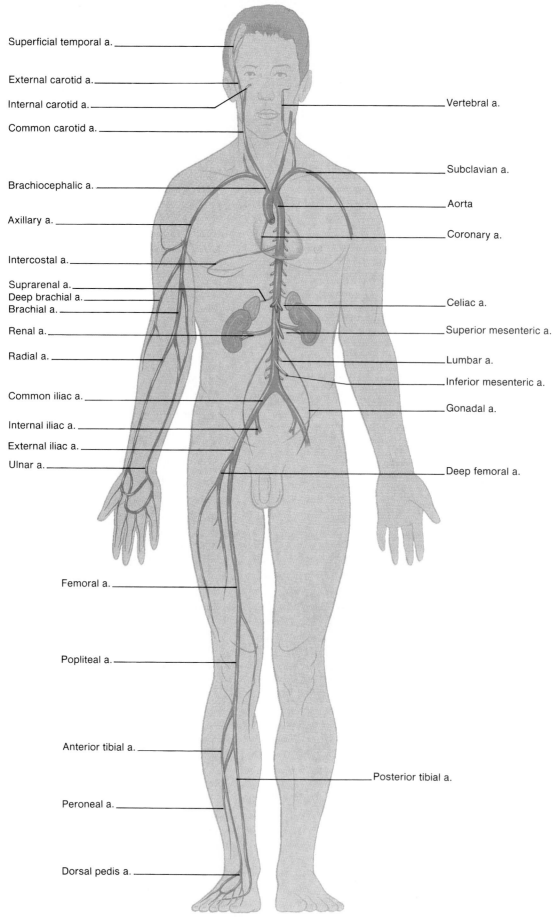

Superficial temporal a.

External carotid a.

Internal carotid a.

Common carotid a.

Vertebral a.

Subclavian a.

Brachiocephalic a.

Aorta

Axillary a.

Coronary a.

Intercostal a.

Suprarenal a.

Deep brachial a.

Brachial a.

Celiac a.

Renal a.

Superior mesenteric a.

Radial a.

Lumbar a.

Inferior mesenteric a.

Common iliac a.

Gonadal a.

Internal iliac a.

External iliac a.

Ulnar a.

Deep femoral a.

Femoral a.

Popliteal a.

Anterior tibial a.

Posterior tibial a.

Peroneal a.

Dorsal pedis a.

Arterial System
Figure 15.52

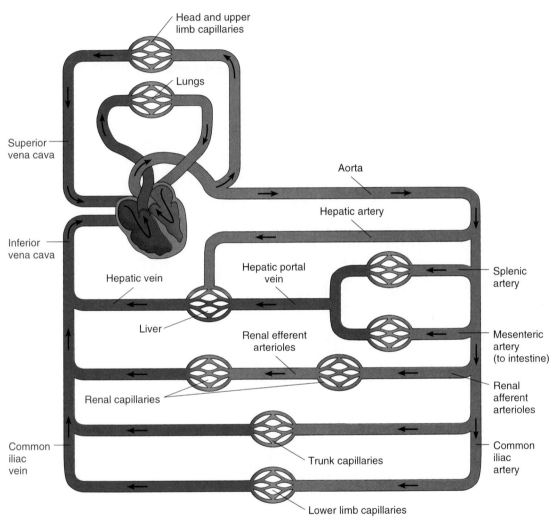

Circulatory System
Figure 15.57

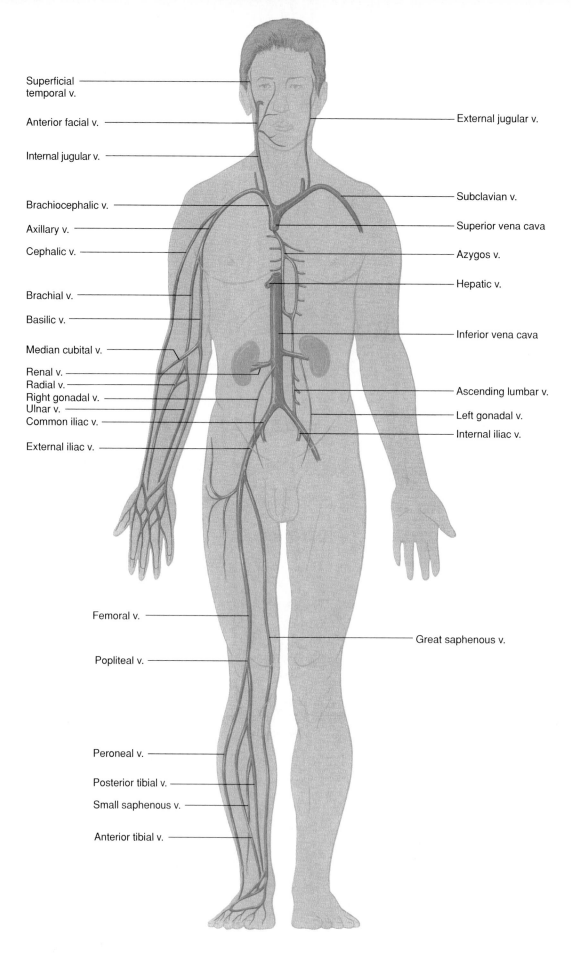

Superficial temporal v.

Anterior facial v.

Internal jugular v.

Brachiocephalic v.

Axillary v.

Cephalic v.

Brachial v.

Basilic v.

Median cubital v.

Renal v.

Radial v.

Right gonadal v.

Ulnar v.

Common iliac v.

External iliac v.

External jugular v.

Subclavian v.

Superior vena cava

Azygos v.

Hepatic v.

Inferior vena cava

Ascending lumbar v.

Left gonadal v.

Internal iliac v.

Femoral v.

Great saphenous v.

Popliteal v.

Peroneal v.

Posterior tibial v.

Small saphenous v.

Anterior tibial v.

Venous System
Figure 15.59

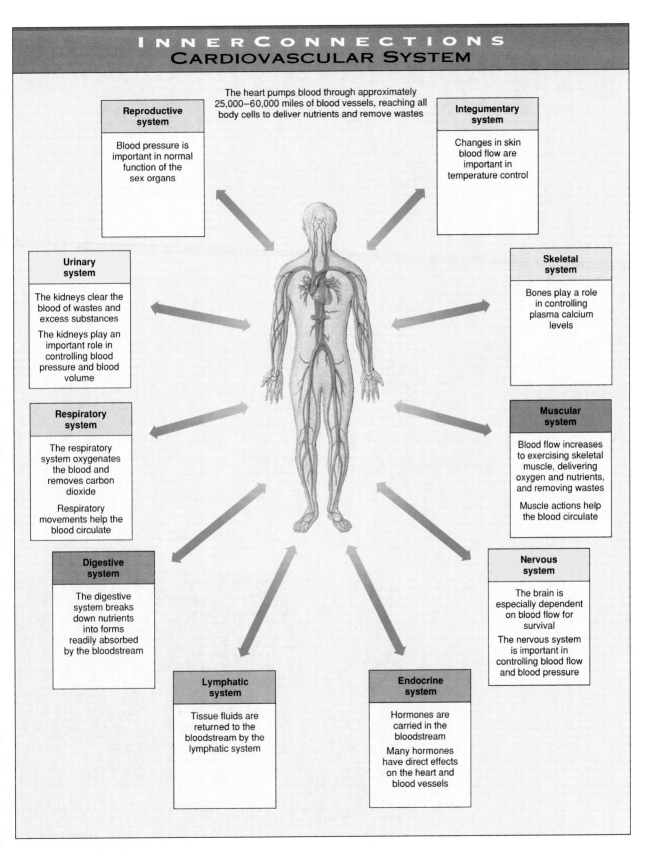

InnerConnections
CARDIOVASCULAR SYSTEM

The heart pumps blood through approximately 25,000–60,000 miles of blood vessels, reaching all body cells to deliver nutrients and remove wastes

Reproductive system

Blood pressure is important in normal function of the sex organs

Integumentary system

Changes in skin blood flow are important in temperature control

Urinary system

The kidneys clear the blood of wastes and excess substances

The kidneys play an important role in controlling blood pressure and blood volume

Skeletal system

Bones play a role in controlling plasma calcium levels

Respiratory system

The respiratory system oxygenates the blood and removes carbon dioxide

Respiratory movements help the blood circulate

Muscular system

Blood flow increases to exercising skeletal muscle, delivering oxygen and nutrients, and removing wastes

Muscle actions help the blood circulate

Digestive system

The digestive system breaks down nutrients into forms readily absorbed by the bloodstream

Nervous system

The brain is especially dependent on blood flow for survival

The nervous system is important in controlling blood flow and blood pressure

Lymphatic system

Tissue fluids are returned to the bloodstream by the lymphatic system

Endocrine system

Hormones are carried in the bloodstream

Many hormones have direct effects on the heart and blood vessels

Cardiovascular System
InnerConnections: Chapter 15

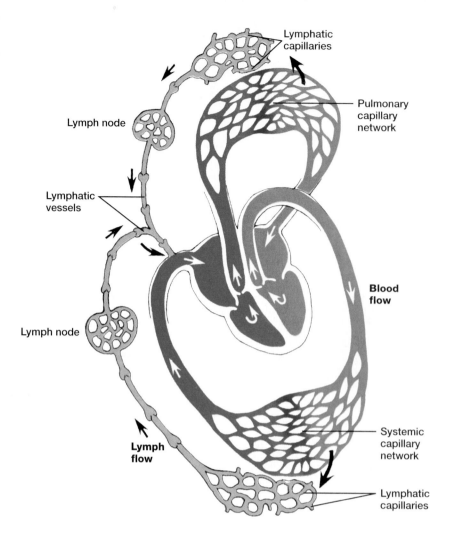

Lymphatic Vessels
Figure 16.1

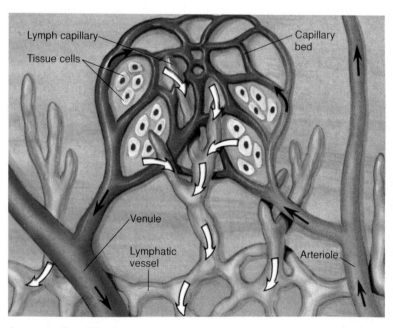

Lymph Capillaries
Figure 16.2

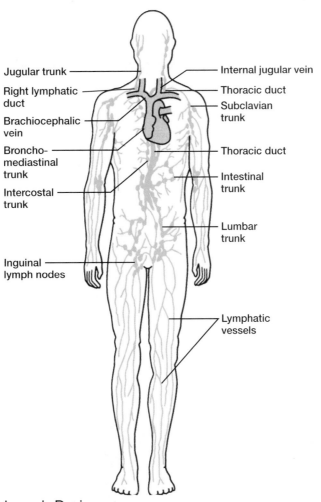

Jugular trunk

Right lymphatic duct

Brachiocephalic vein

Broncho-mediastinal trunk

Intercostal trunk

Inguinal lymph nodes

Internal jugular vein

Thoracic duct

Subclavian trunk

Thoracic duct

Intestinal trunk

Lumbar trunk

Lymphatic vessels

Lymph Drainage
Figure 16.4

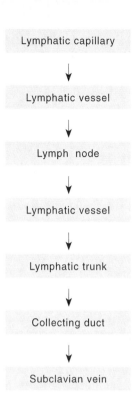

Lymphatic capillary

↓

Lymphatic vessel

↓

Lymph node

↓

Lymphatic vessel

↓

Lymphatic trunk

↓

Collecting duct

↓

Subclavian vein

The Lymphatic Pathway
Figure 16.7

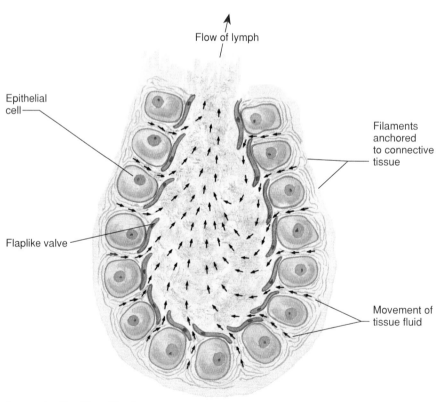

Flow of lymph

Epithelial cell

Filaments anchored to connective tissue

Flaplike valve

Movement of tissue fluid

Lymphatic Capillaries
Figure 16.8

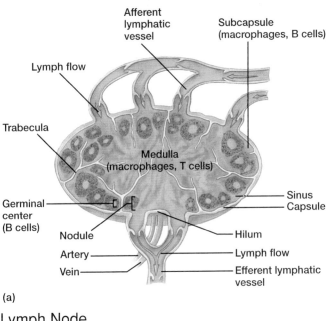

Lymph Node
Figure 16.9a

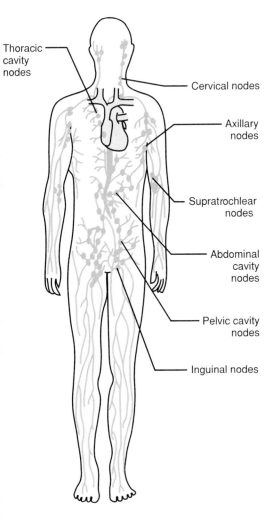

Lymph Node Locations
Figure 16.11

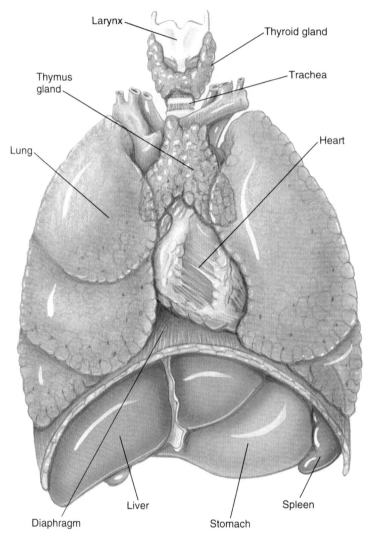

Larynx

Thyroid gland

Thymus gland

Trachea

Lung

Heart

Diaphragm

Liver

Stomach

Spleen

The Thymus Gland
Figure 16.12

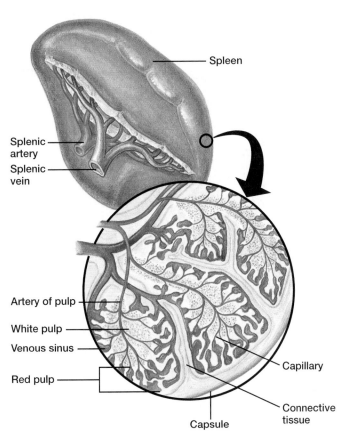

Spleen

Splenic artery

Splenic vein

Artery of pulp

White pulp

Venous sinus

Red pulp

Capillary

Connective tissue

Capsule

The Spleen
Figure 16.14

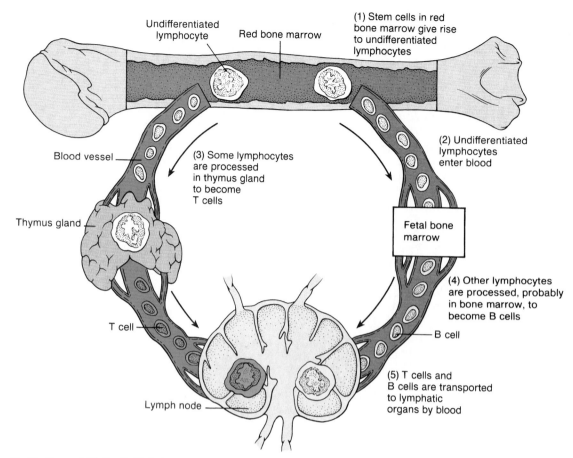

Undifferentiated lymphocyte

Red bone marrow

(1) Stem cells in red bone marrow give rise to undifferentiated lymphocytes

(2) Undifferentiated lymphocytes enter blood

Blood vessel

(3) Some lymphocytes are processed in thymus gland to become T cells

Thymus gland

Fetal bone marrow

(4) Other lymphocytes are processed, probably in bone marrow, to become B cells

T cell

B cell

(5) T cells and B cells are transported to lymphatic organs by blood

Lymph node

B-Cell and T-Cell Origins
Figure 16.16

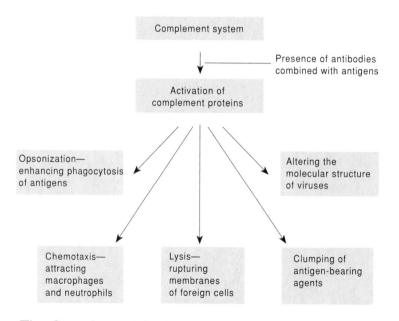

Complement system

Presence of antibodies combined with antigens

Activation of complement proteins

Opsonization— enhancing phagocytosis of antigens

Altering the molecular structure of viruses

Chemotaxis— attracting macrophages and neutrophils

Lysis— rupturing membranes of foreign cells

Clumping of antigen-bearing agents

The Complement System
Figure 16.19

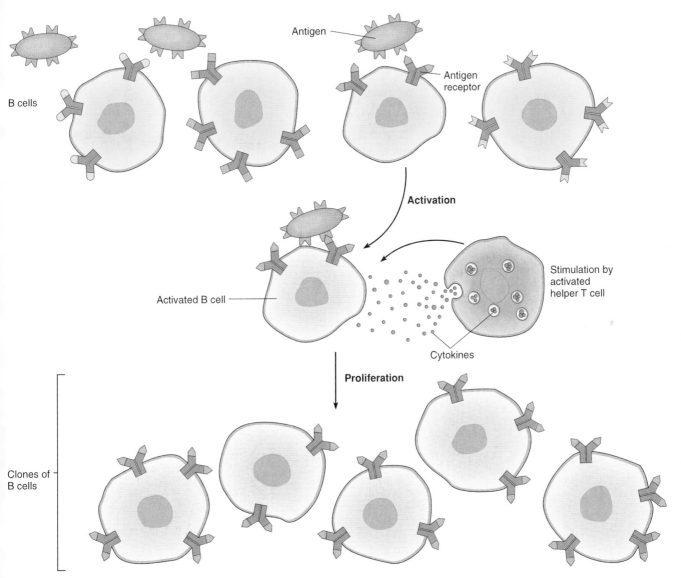

B cells

Antigen

Antigen
receptor

Activation

Activated B cell

Stimulation by
activated
helper T cell

Cytokines

Proliferation

Clones of
B cells

B-Cell Activation
Figure 16.20

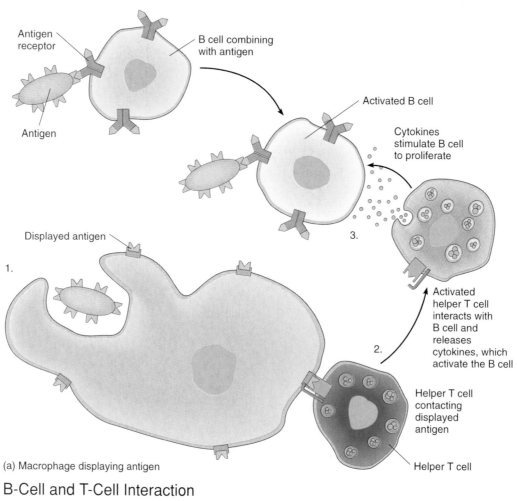

Antigen receptor

B cell combining with antigen

Antigen

Activated B cell

Cytokines stimulate B cell to proliferate

3.

Displayed antigen

1.

Activated helper T cell interacts with B cell and releases cytokines, which activate the B cell

2.

Helper T cell contacting displayed antigen

Helper T cell

(a) Macrophage displaying antigen

B-Cell and T-Cell Interaction
Figure 16.21a

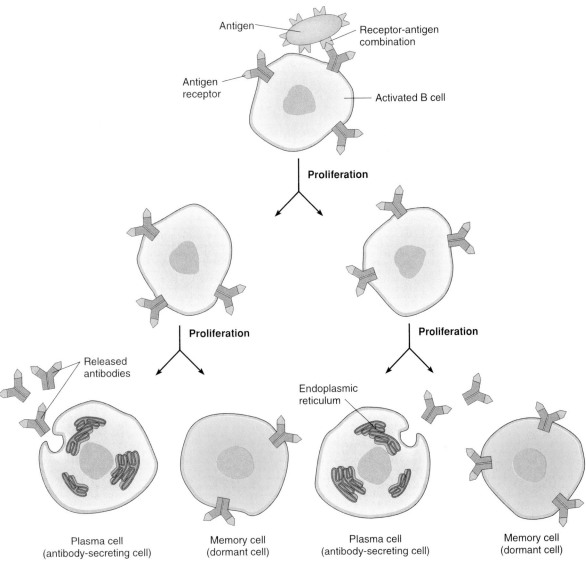

Antigen

Receptor-antigen combination

Antigen receptor

Activated B cell

Proliferation

Proliferation

Proliferation

Released antibodies

Endoplasmic reticulum

Plasma cell
(antibody-secreting cell)

Memory cell
(dormant cell)

Plasma cell
(antibody-secreting cell)

Memory cell
(dormant cell)

B-Cell Proliferation
Figure 16.22

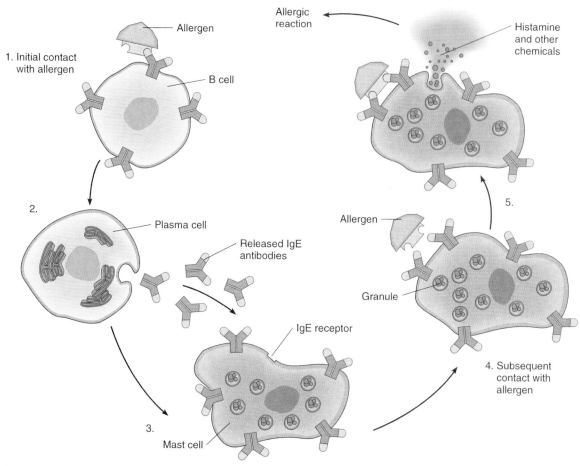

Immediate-Reaction Allergy
Figure 16.24a

Allergen

1. Initial contact with allergen

B cell

2.

Plasma cell

Released IgE antibodies

IgE receptor

3.

Mast cell

Allergic reaction

Histamine and other chemicals

5.

Allergen

Granule

4. Subsequent contact with allergen

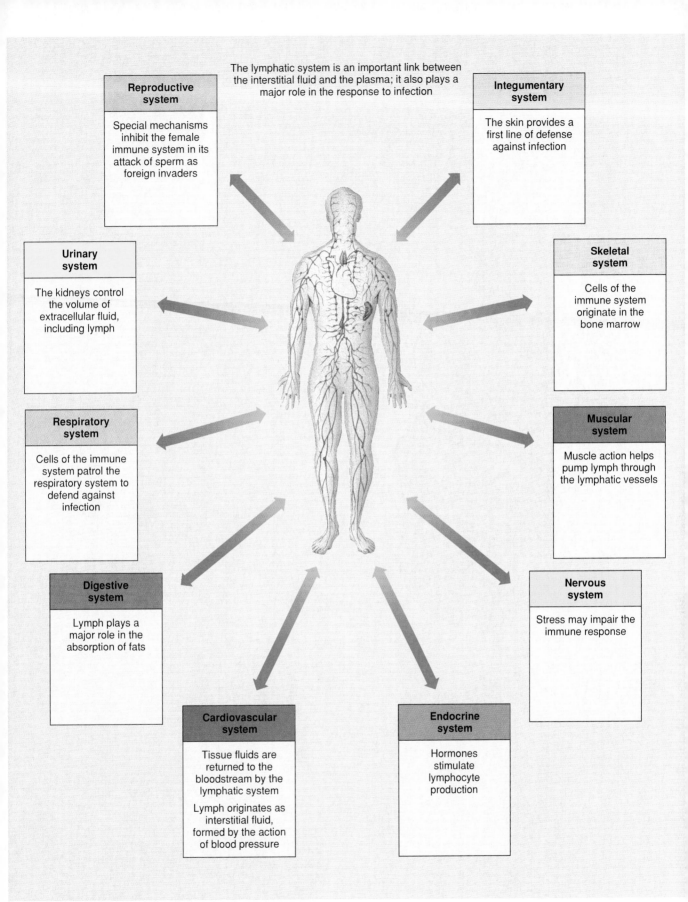

The lymphatic system is an important link between the interstitial fluid and the plasma; it also plays a major role in the response to infection

Reproductive system

Special mechanisms inhibit the female immune system in its attack of sperm as foreign invaders

Integumentary system

The skin provides a first line of defense against infection

Urinary system

The kidneys control the volume of extracellular fluid, including lymph

Skeletal system

Cells of the immune system originate in the bone marrow

Respiratory system

Cells of the immune system patrol the respiratory system to defend against infection

Muscular system

Muscle action helps pump lymph through the lymphatic vessels

Digestive system

Lymph plays a major role in the absorption of fats

Nervous system

Stress may impair the immune response

Cardiovascular system

Tissue fluids are returned to the bloodstream by the lymphatic system

Lymph originates as interstitial fluid, formed by the action of blood pressure

Endocrine system

Hormones stimulate lymphocyte production

Lymphatic System
InnerConnections: Chapter 16

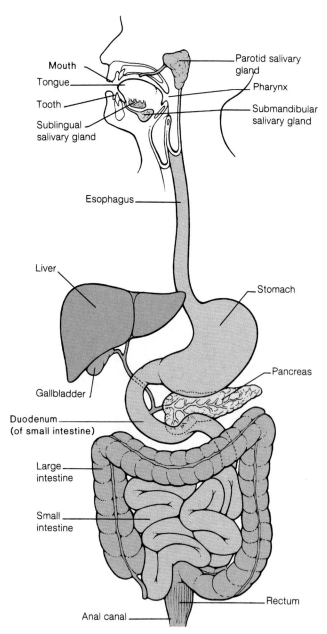

Mouth
Tongue
Tooth
Sublingual
salivary gland
Parotid salivary
gland
Pharynx
Submandibular
salivary gland
Esophagus
Liver
Stomach
Gallbladder
Pancreas
Duodenum
(of small intestine)
Large
intestine
Small
intestine
Rectum
Anal canal

Digestive System
Figure 17.1

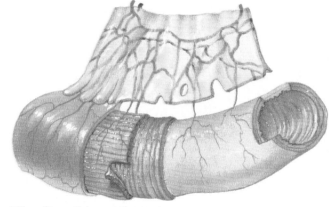

The Small Intestine
Figure 17.3

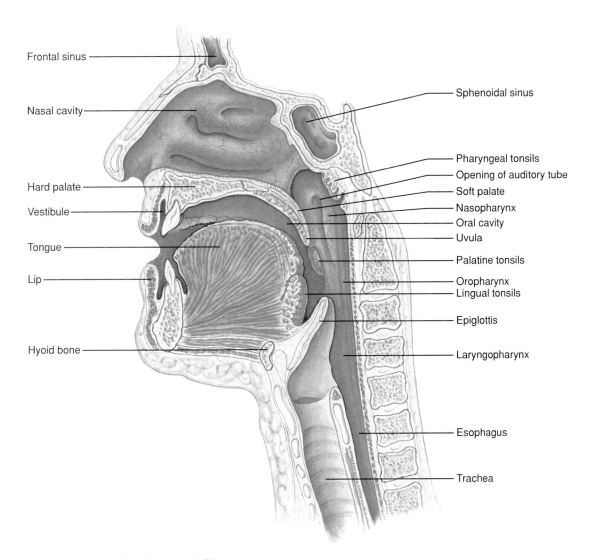

Frontal sinus

Nasal cavity

Hard palate

Vestibule

Tongue

Lip

Hyoid bone

Sphenoidal sinus

Pharyngeal tonsils

Opening of auditory tube

Soft palate

Nasopharynx

Oral cavity

Uvula

Palatine tonsils

Oropharynx

Lingual tonsils

Epiglottis

Laryngopharynx

Esophagus

Trachea

Mouth, Nasal Cavity, and Pharynx
Figure 17.7

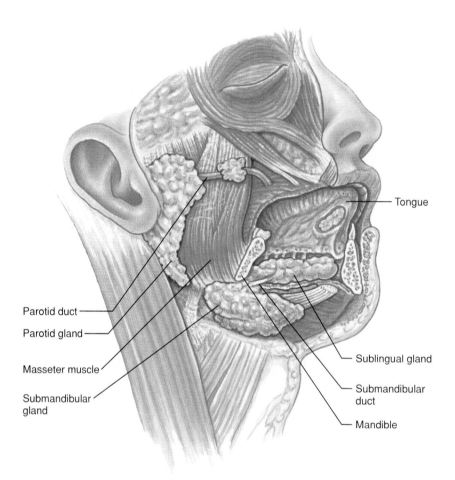

Parotid duct

Parotid gland

Masseter muscle

Submandibular
gland

Tongue

Sublingual gland

Submandibular
duct

Mandible

Major Salivary Glands
Figure 17.11

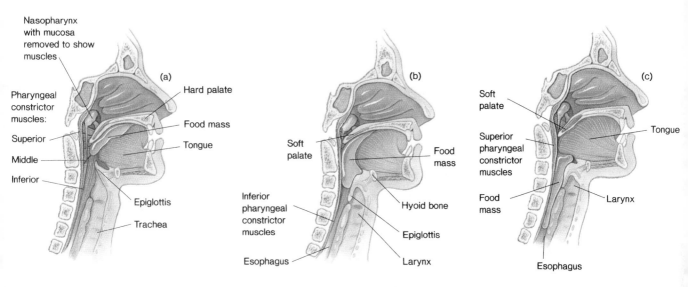

(a)

Nasopharynx with mucosa removed to show muscles

Pharyngeal constrictor muscles:

Superior

Middle

Inferior

Hard palate

Food mass

Tongue

Epiglottis

Trachea

(b)

Soft palate

Inferior pharyngeal constrictor muscles

Esophagus

Food mass

Hyoid bone

Epiglottis

Larynx

(c)

Soft palate

Superior pharyngeal constrictor muscles

Food mass

Tongue

Larynx

Esophagus

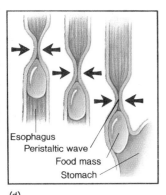

Esophagus
Peristaltic wave
Food mass
Stomach

(d)

Swallowing

Figure 17.14

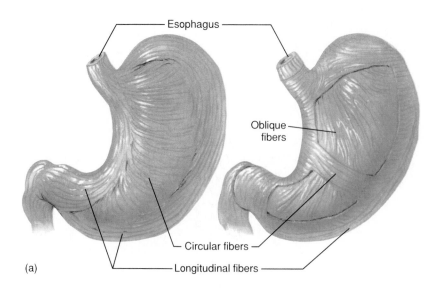

Esophagus

Oblique
fibers

Circular fibers

(a) Longitudinal fibers

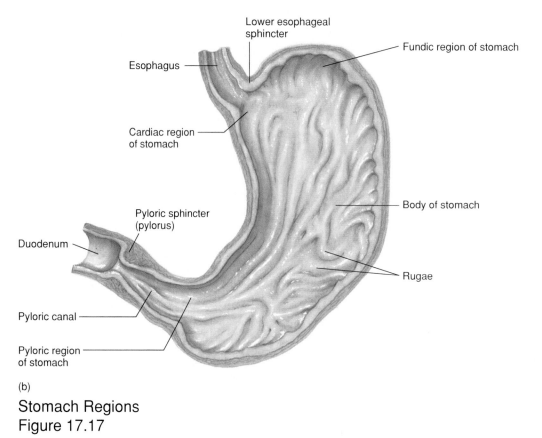

Lower esophageal
sphincter

Fundic region of stomach

Esophagus

Cardiac region
of stomach

Body of stomach

Pyloric sphincter
(pylorus)

Duodenum

Rugae

Pyloric canal

Pyloric region
of stomach

(b)

Stomach Regions
Figure 17.17

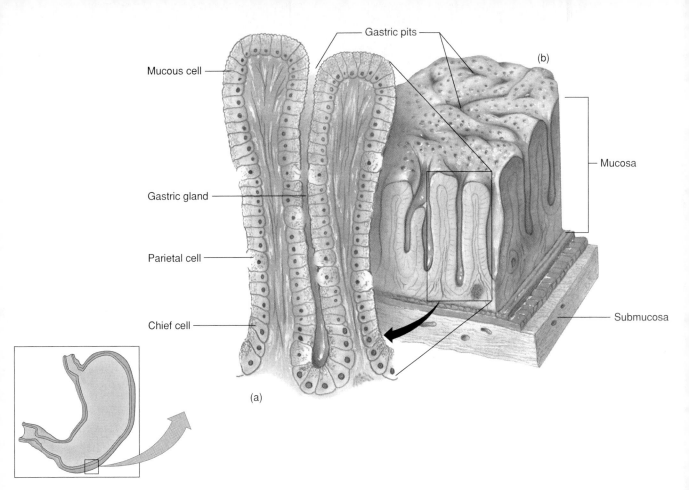

- Mucous cell
- Gastric gland
- Parietal cell
- Chief cell
- Gastric pits
- (b)
- Mucosa
- Submucosa
- (a)

Gastric Glands
Figure 17.19

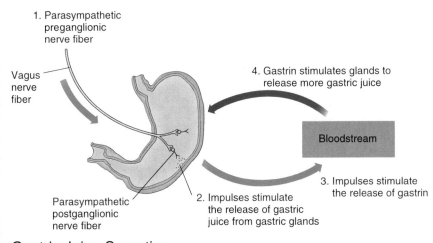

1. Parasympathetic preganglionic nerve fiber

Vagus nerve fiber

4. Gastrin stimulates glands to release more gastric juice

Bloodstream

3. Impulses stimulate the release of gastrin

Parasympathetic postganglionic nerve fiber

2. Impulses stimulate the release of gastric juice from gastric glands

Gastric Juice Secretion
Figure 17.21

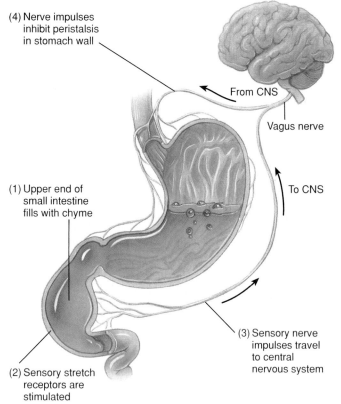

(4) Nerve impulses inhibit peristalsis in stomach wall

From CNS

Vagus nerve

To CNS

(1) Upper end of small intestine fills with chyme

(3) Sensory nerve impulses travel to central nervous system

(2) Sensory stretch receptors are stimulated

The Enterogastric Reflex
Figure 17.23

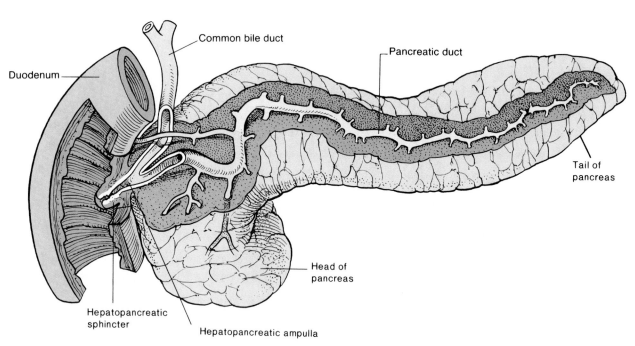

Common bile duct

Pancreatic duct

Duodenum

Tail of pancreas

Head of pancreas

Hepatopancreatic sphincter

Hepatopancreatic ampulla

The Pancreas
Figure 17.24

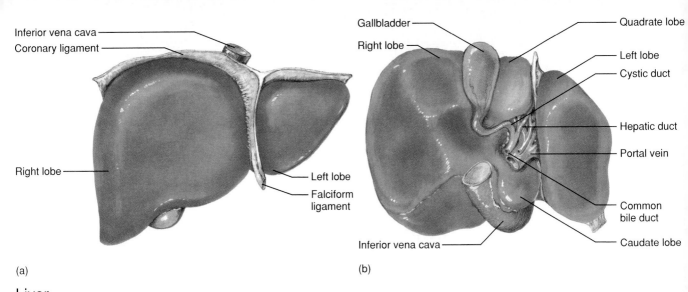

(a)

(b)

Inferior vena cava
Coronary ligament
Right lobe
Left lobe
Falciform ligament

Gallbladder
Right lobe
Quadrate lobe
Left lobe
Cystic duct
Hepatic duct
Portal vein
Common bile duct
Caudate lobe
Inferior vena cava

Liver
Figure 17.28

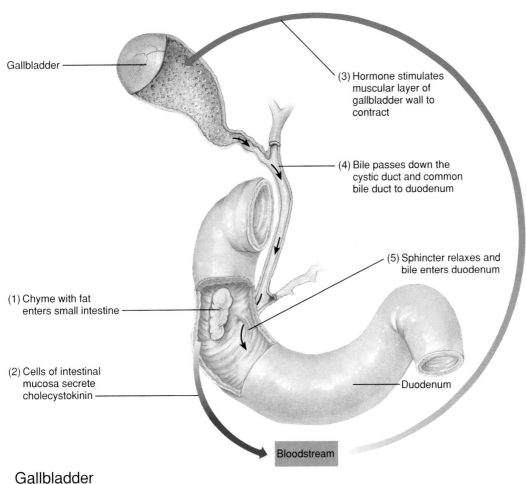

Gallbladder

(3) Hormone stimulates muscular layer of gallbladder wall to contract

(4) Bile passes down the cystic duct and common bile duct to duodenum

(5) Sphincter relaxes and bile enters duodenum

(1) Chyme with fat enters small intestine

(2) Cells of intestinal mucosa secrete cholecystokinin

Duodenum

Bloodstream

Gallbladder
Figure 17.32

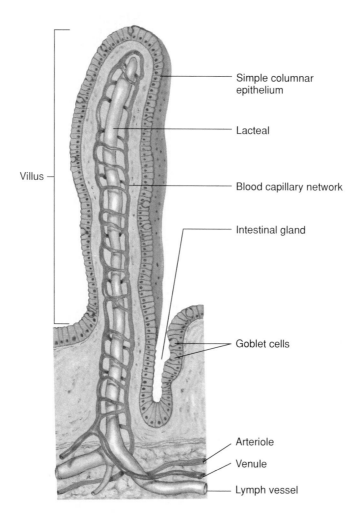

Simple columnar epithelium

Lacteal

Blood capillary network

Intestinal gland

Goblet cells

Arteriole

Venule

Lymph vessel

Villus

Intestinal Villus
Figure 17.37

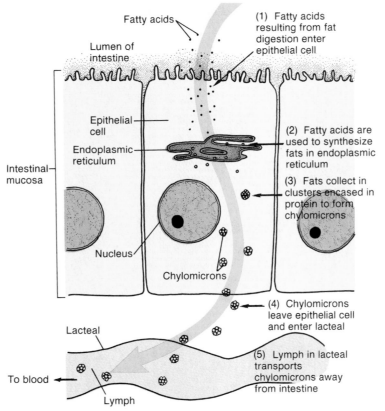

Fatty acids

Lumen of
intestine

(1) Fatty acids
resulting from fat
digestion enter
epithelial cell

Epithelial
cell

Endoplasmic
reticulum

Intestinal
mucosa

(2) Fatty acids are
used to synthesize
fats in endoplasmic
reticulum

(3) Fats collect in
clusters encased in
protein to form
chylomicrons

Nucleus

Chylomicrons

(4) Chylomicrons
leave epithelial cell
and enter lacteal

Lacteal

To blood

(5) Lymph in lacteal
transports
chylomicrons away
from intestine

Lymph

Fatty Acid Absorption
Figure 17.44

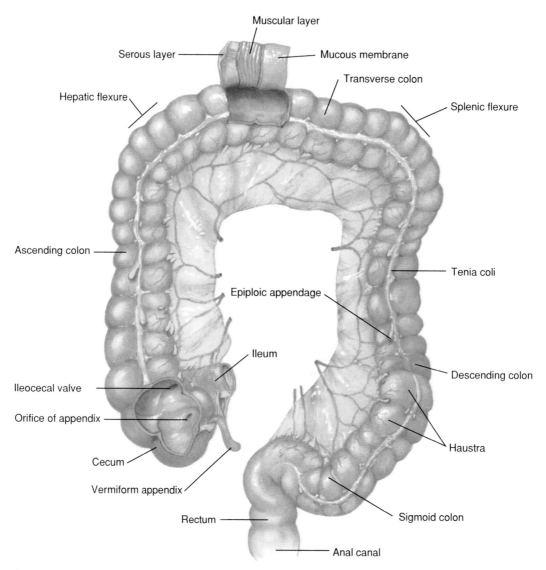

Muscular layer

Serous layer

Mucous membrane

Transverse colon

Hepatic flexure

Splenic flexure

Ascending colon

Tenia coli

Epiploic appendage

Descending colon

Ileum

Ileocecal valve

Orifice of appendix

Haustra

Cecum

Vermiform appendix

Rectum

Sigmoid colon

Anal canal

Large Intestine
Figure 17.45

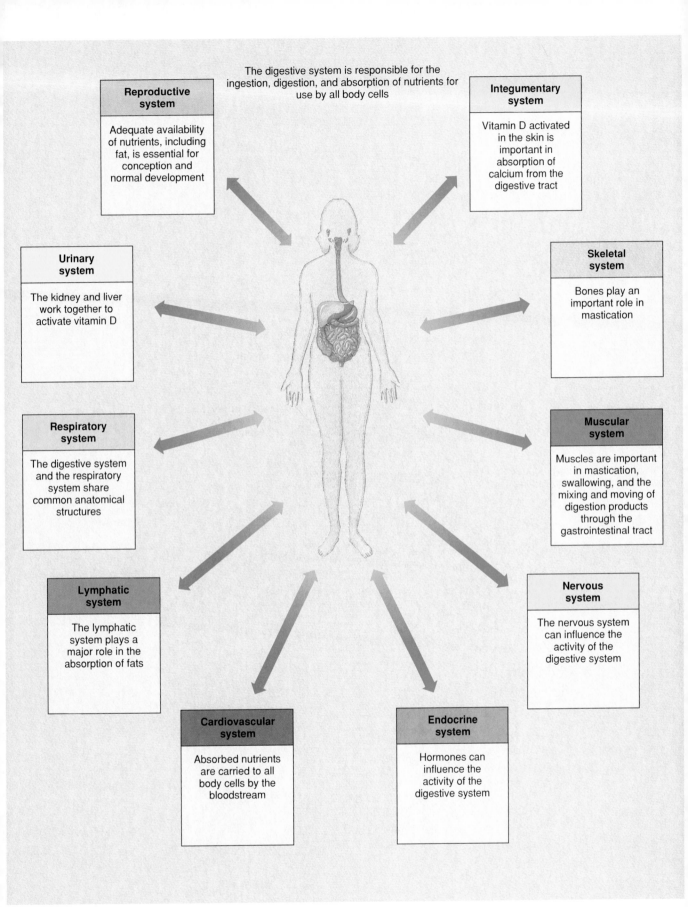

The digestive system is responsible for the ingestion, digestion, and absorption of nutrients for use by all body cells

Reproductive system

Adequate availability of nutrients, including fat, is essential for conception and normal development

Integumentary system

Vitamin D activated in the skin is important in absorption of calcium from the digestive tract

Urinary system

The kidney and liver work together to activate vitamin D

Skeletal system

Bones play an important role in mastication

Respiratory system

The digestive system and the respiratory system share common anatomical structures

Muscular system

Muscles are important in mastication, swallowing, and the mixing and moving of digestion products through the gastrointestinal tract

Lymphatic system

The lymphatic system plays a major role in the absorption of fats

Nervous system

The nervous system can influence the activity of the digestive system

Cardiovascular system

Absorbed nutrients are carried to all body cells by the bloodstream

Endocrine system

Hormones can influence the activity of the digestive system

Digestive System
InnerConnections: Chapter 17

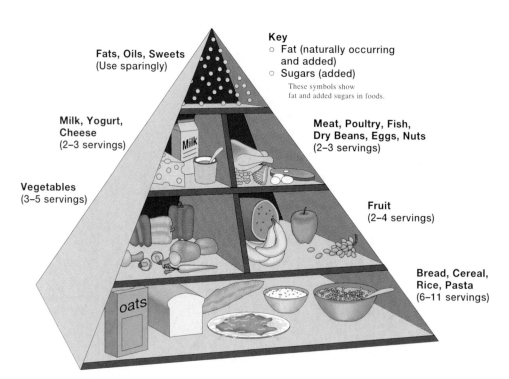

Fats, Oils, Sweets
(Use sparingly)

Key
○ Fat (naturally occurring
   and added)
○ Sugars (added)
These symbols show
fat and added sugars in foods.

Milk, Yogurt,
Cheese
(2–3 servings)

Meat, Poultry, Fish,
Dry Beans, Eggs, Nuts
(2–3 servings)

Vegetables
(3–5 servings)

Fruit
(2–4 servings)

Bread, Cereal,
Rice, Pasta
(6–11 servings)

Food Pyramid
Figure 18.16

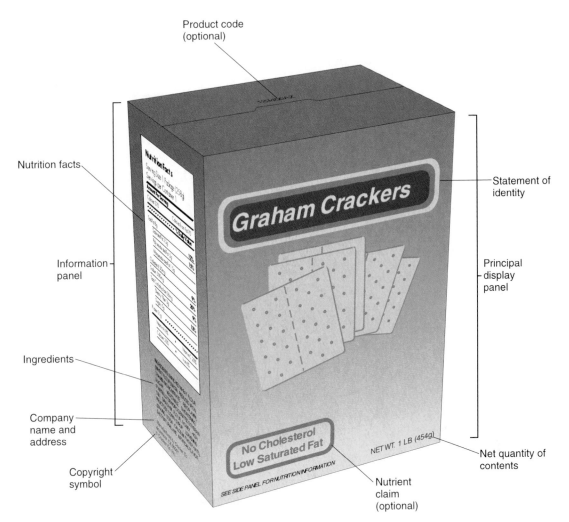

Product code
(optional)

Nutrition facts

Information
panel

Ingredients

Company
name and
address

Copyright
symbol

Statement of
identity

Principal
display
panel

Net quantity of
contents

Nutrient
claim
(optional)

Graham Crackers

No Cholesterol
Low Saturated Fat

NET WT. 1 LB (454g)

SEE SIDE PANEL FOR NUTRITION INFORMATION

"Nutrition Facts" Panel
Figure 18.17

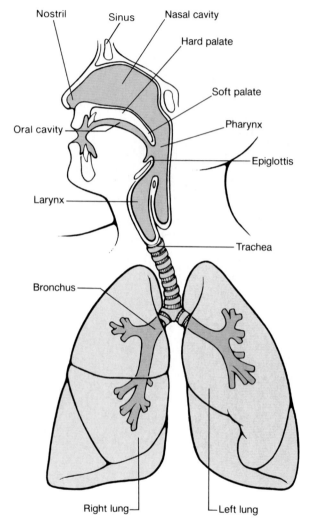

Nostril

Sinus

Nasal cavity

Hard palate

Soft palate

Oral cavity

Pharynx

Epiglottis

Larynx

Trachea

Bronchus

Right lung

Left lung

Respiratory System
Figure 19.1

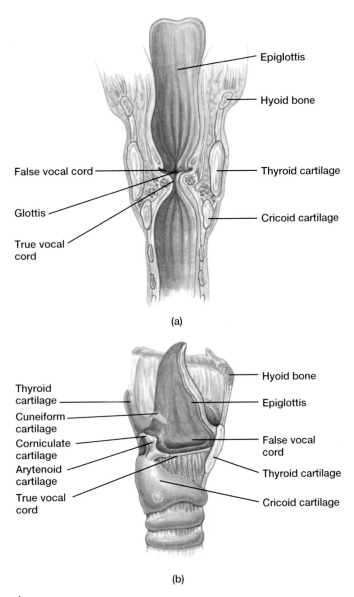

(a)

(b)

Larynx
Figure 19.6

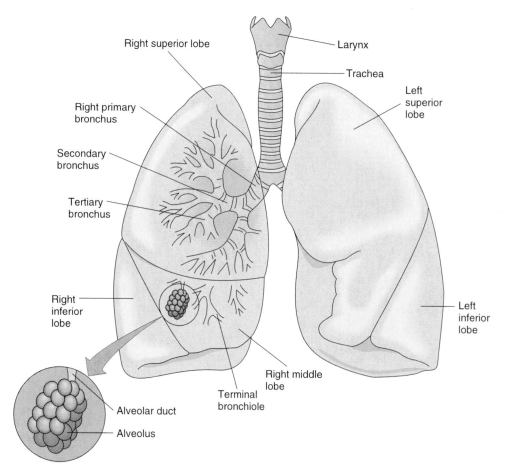

Right superior lobe

Larynx

Trachea

Left superior lobe

Right primary bronchus

Secondary bronchus

Tertiary bronchus

Right inferior lobe

Left inferior lobe

Right middle lobe

Terminal bronchiole

Alveolar duct

Alveolus

Bronchial Tree
Figure 19.12

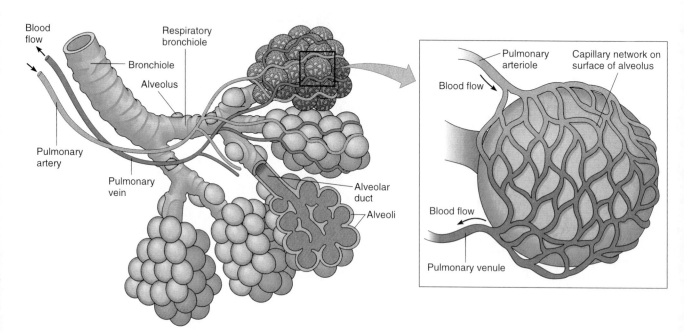

Blood flow

Respiratory bronchiole

Bronchiole

Alveolus

Pulmonary artery

Pulmonary vein

Alveolar duct

Alveoli

Pulmonary arteriole

Capillary network on surface of alveolus

Blood flow

Blood flow

Pulmonary venule

Respiratory Tubes and Alveoli
Figure 19.14

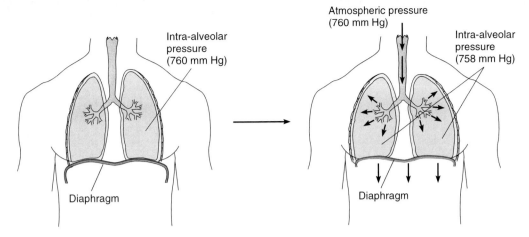

Intra-Alveolar Pressure Changes
Figure 19.22

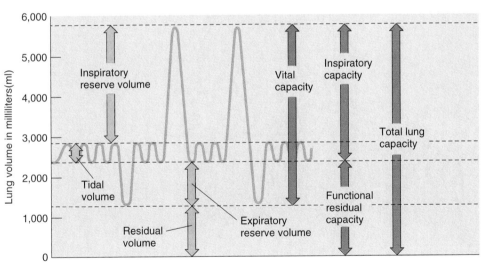

Lung Volumes
Figure 19.25

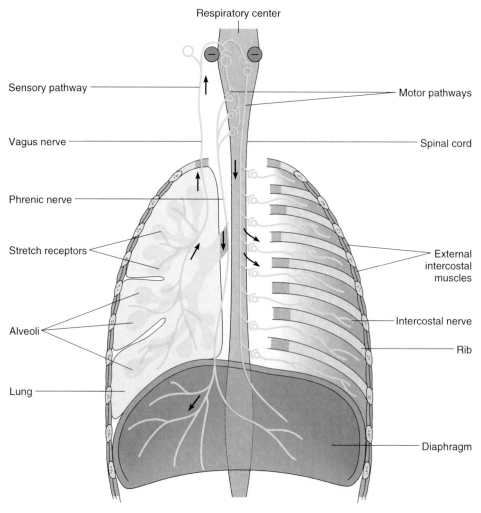

Respiratory center

Sensory pathway

Motor pathways

Vagus nerve

Spinal cord

Phrenic nerve

Stretch receptors

External intercostal muscles

Intercostal nerve

Alveoli

Rib

Lung

Diaphragm

Inspiration
Figure 19.30

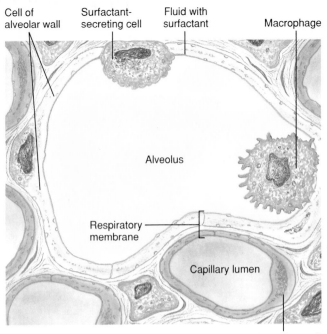

Cell of alveolar wall

Surfactant-secreting cell

Fluid with surfactant

Macrophage

Alveolus

Respiratory membrane

Capillary lumen

Cell of capillary wall

Respiratory Membrane
Figure 19.32

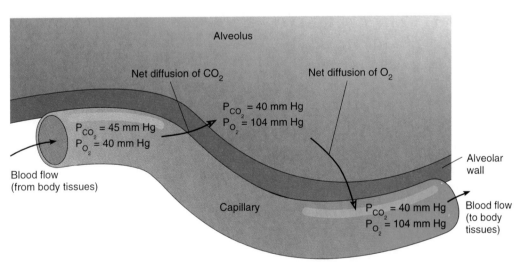

Alveolus

Net diffusion of CO$_2$

Net diffusion of O$_2$

$P_{CO_2} = 45$ mm Hg
$P_{O_2} = 40$ mm Hg

$P_{CO_2} = 40$ mm Hg
$P_{O_2} = 104$ mm Hg

Blood flow (from body tissues)

Capillary

Alveolar wall

$P_{CO_2} = 40$ mm Hg
$P_{O_2} = 104$ mm Hg

Blood flow (to body tissues)

Alveolar Gas Exchange
Figure 19.34

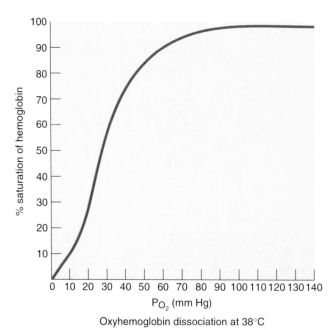

Oxyhemoglobin dissociation at 38°C

Oxyhemoglobin Dissociation Curve
Figure 19.35

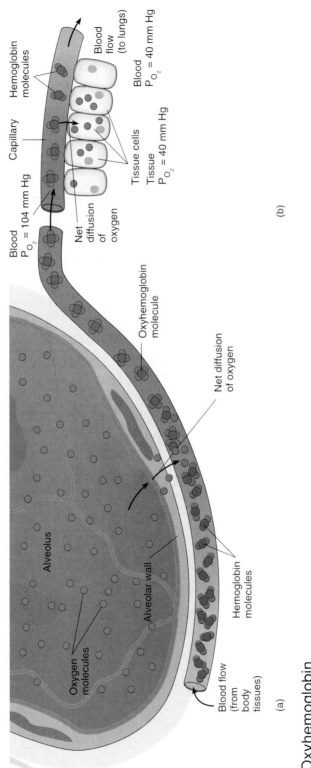

Blood flow (from body tissues)

Blood flow (to lungs)

Alveolus

Oxygen molecules

Alveolar wall

Hemoglobin molecules

Net diffusion of oxygen

Oxyhemoglobin molecule

Blood P_{O_2} = 104 mm Hg

Net diffusion of oxygen

Capillary

Hemoglobin molecules

Blood P_{O_2} = 40 mm Hg

Tissue cells

Tissue P_{O_2} = 40 mm Hg

Blood P_{O_2} = 40 mm Hg

(a)

(b)

Oxyhemoglobin
Figure 19.36

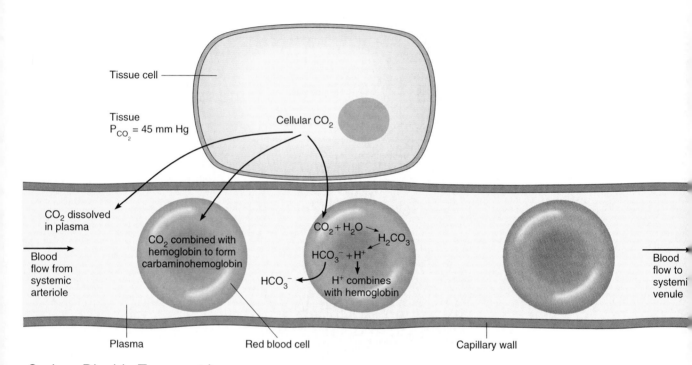

Tissue cell

Tissue P_{CO_2} = 45 mm Hg

Cellular CO_2

CO_2 dissolved in plasma

Blood flow from systemic arteriole

CO_2 combined with hemoglobin to form carbaminohemoglobin

$CO_2 + H_2O \longrightarrow H_2CO_3$

$HCO_3^- + H^+$

HCO_3^-

H^+ combines with hemoglobin

Blood flow to systemic venule

Plasma

Red blood cell

Capillary wall

Carbon Dioxide Transport I
Figure 19.40

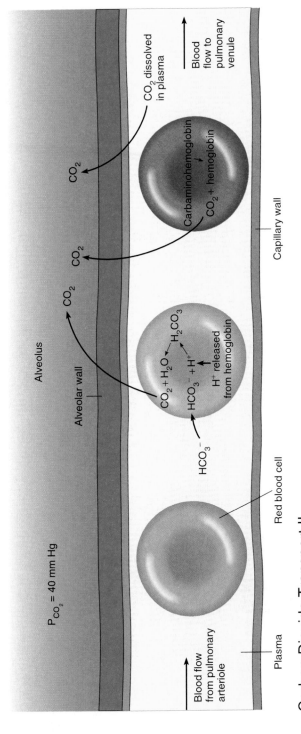

Carbon Dioxide Transport II
Figure 19.42

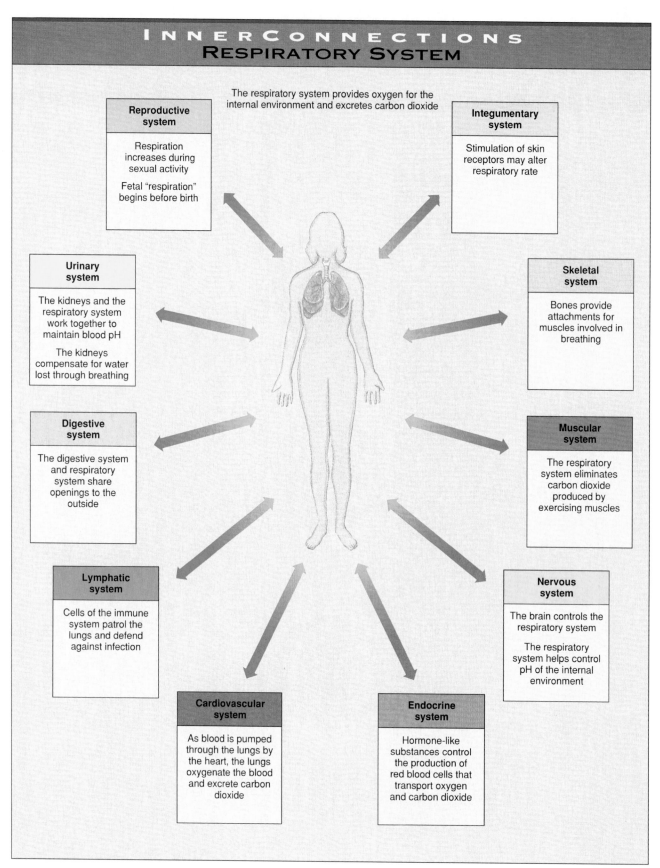

INNERCONNECTIONS
RESPIRATORY SYSTEM

The respiratory system provides oxygen for the internal environment and excretes carbon dioxide

Reproductive system

Respiration increases during sexual activity

Fetal "respiration" begins before birth

Integumentary system

Stimulation of skin receptors may alter respiratory rate

Urinary system

The kidneys and the respiratory system work together to maintain blood pH

The kidneys compensate for water lost through breathing

Skeletal system

Bones provide attachments for muscles involved in breathing

Digestive system

The digestive system and respiratory system share openings to the outside

Muscular system

The respiratory system eliminates carbon dioxide produced by exercising muscles

Lymphatic system

Cells of the immune system patrol the lungs and defend against infection

Nervous system

The brain controls the respiratory system

The respiratory system helps control pH of the internal environment

Cardiovascular system

As blood is pumped through the lungs by the heart, the lungs oxygenate the blood and excrete carbon dioxide

Endocrine system

Hormone-like substances control the production of red blood cells that transport oxygen and carbon dioxide

Respiratory System
InnerConnections: Chapter 19

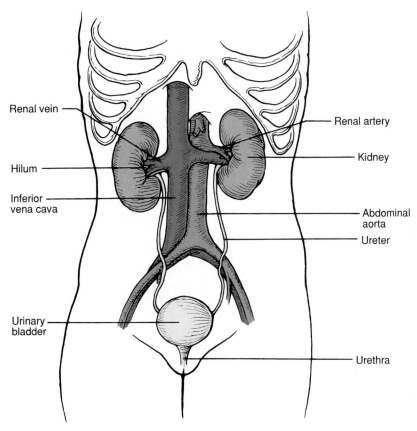

Renal vein

Hilum

Inferior
vena cava

Renal artery

Kidney

Abdominal
aorta

Ureter

Urinary
bladder

Urethra

Urinary System
Figure 20.1

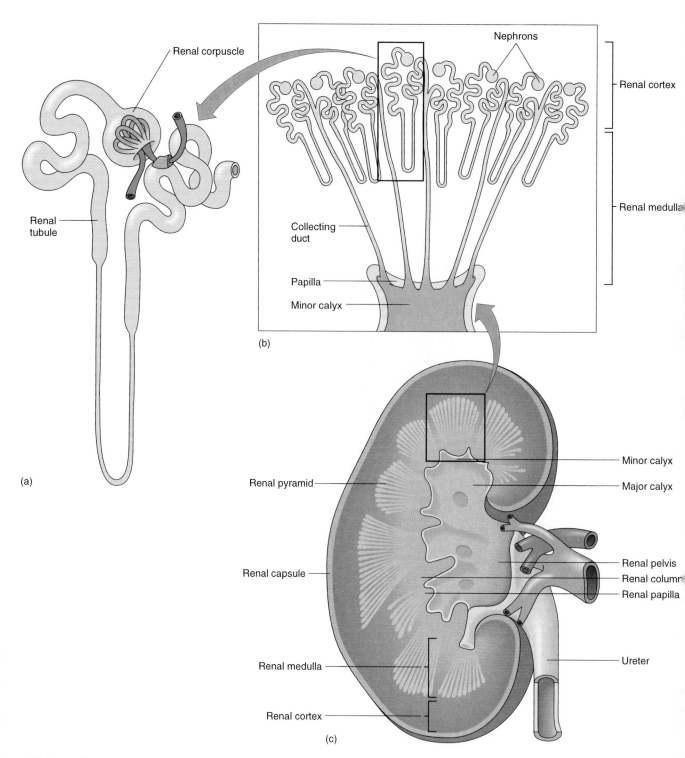

Renal corpuscle

Renal tubule

(a)

Nephrons

Renal cortex

Renal medulla

Collecting duct

Papilla

Minor calyx

(b)

Renal pyramid

Renal capsule

Renal medulla

Renal cortex

(c)

Minor calyx

Major calyx

Renal pelvis

Renal column

Renal papilla

Ureter

Kidney Structure
Figure 20.4

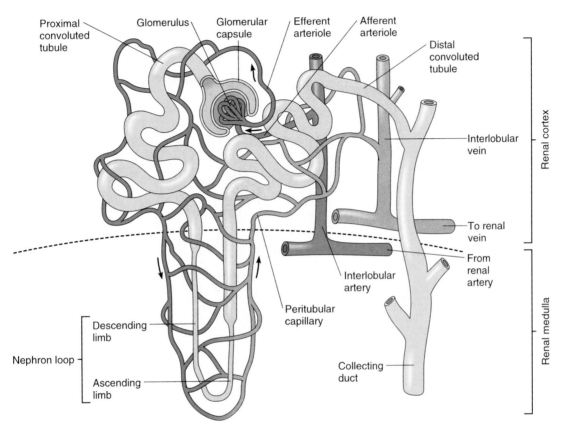

Proximal convoluted tubule
Glomerulus
Glomerular capsule
Efferent arteriole
Afferent arteriole
Distal convoluted tubule
Interlobular vein
To renal vein
From renal artery
Interlobular artery
Peritubular capillary
Descending limb
Ascending limb
Nephron loop
Collecting duct
Renal cortex
Renal medulla

Nephron Structure
Figure 20.9

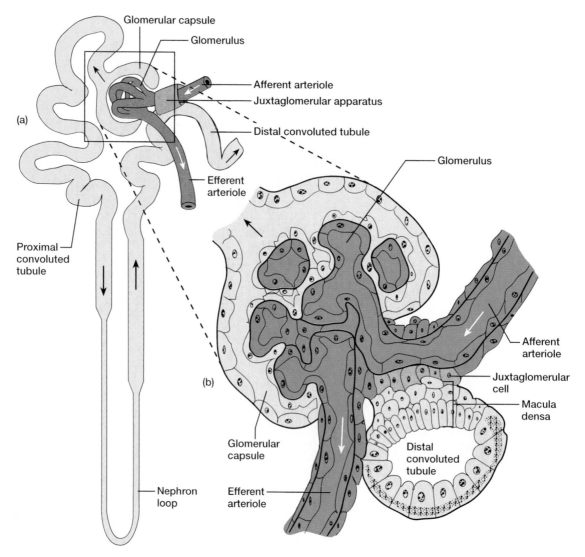

(a)

Glomerular capsule

Glomerulus

Afferent arteriole

Juxtaglomerular apparatus

Distal convoluted tubule

Efferent arteriole

Proximal convoluted tubule

Nephron loop

(b)

Glomerulus

Afferent arteriole

Juxtaglomerular cell

Macula densa

Distal convoluted tubule

Glomerular capsule

Efferent arteriole

Juxtaglomerular Apparatus
Figure 20.11

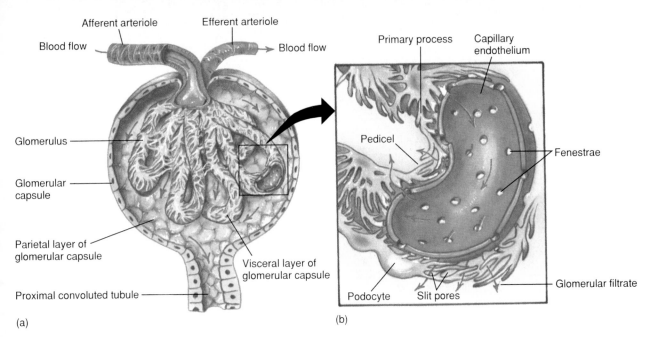

(a)

(b)

Glomerular Filtration

Figure 20.15

Afferent arteriole

Efferent arteriole

Blood flow

Blood flow

Glomerulus

Glomerular capsule

Parietal layer of glomerular capsule

Visceral layer of glomerular capsule

Proximal convoluted tubule

Primary process

Capillary endothelium

Pedicel

Fenestrae

Podocyte

Slit pores

Glomerular filtrate

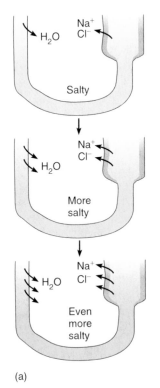

H_2O

Na^+
Cl^-

Salty

H_2O

Na^+
Cl^-

More salty

H_2O

Na^+
Cl^-

Even more salty

(a)

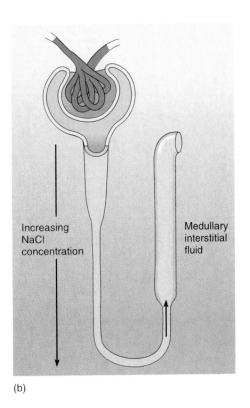

Increasing NaCl concentration

Medullary interstitial fluid

(b)

Tubular Reabsorption

Figure 20.19

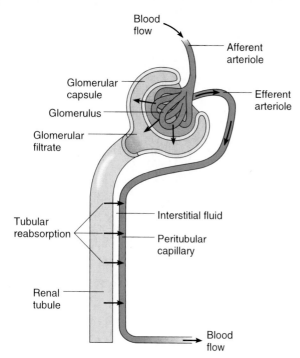

The Countercurrent Multiplier
Figure 20.23

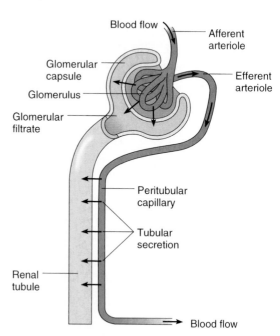

Tubular Secretion
Figure 20.25

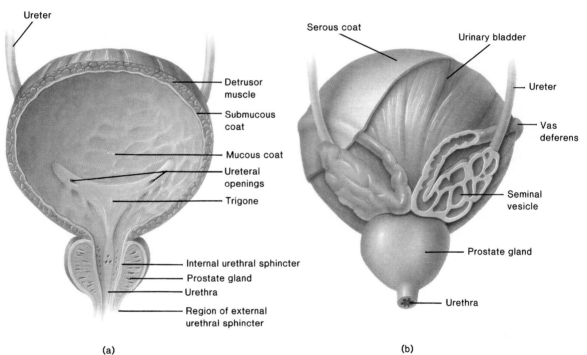

Ureter

Detrusor
muscle

Submucous
coat

Mucous coat

Ureteral
openings

Trigone

Internal urethral sphincter

Prostate gland

Urethra

Region of external
urethral sphincter

(a)

Serous coat

Urinary bladder

Ureter

Vas
deferens

Seminal
vesicle

Prostate gland

Urethra

(b)

Urinary Bladder
Figure 20.29

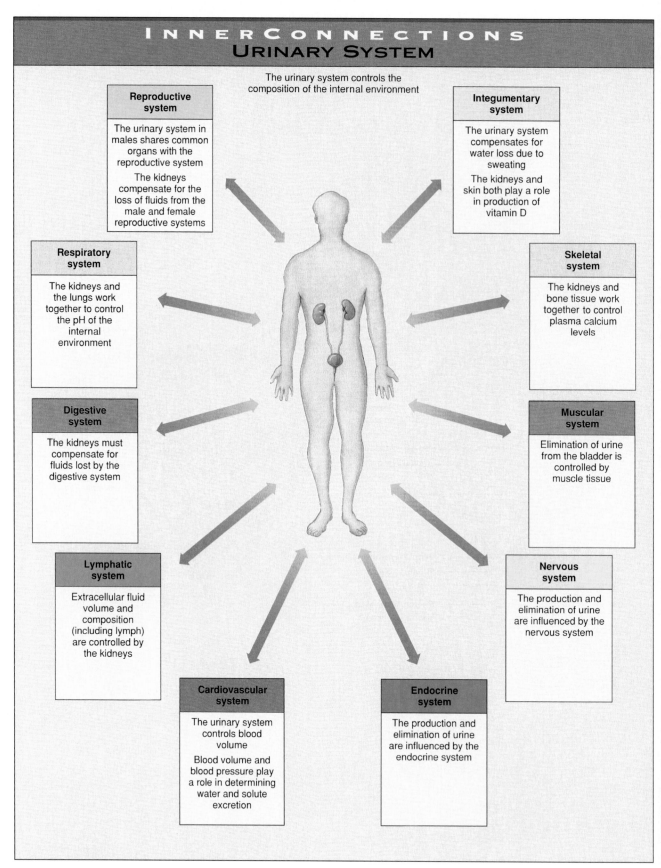

InnerConnections
Urinary System

The urinary system controls the composition of the internal environment

Reproductive system

The urinary system in males shares common organs with the reproductive system

The kidneys compensate for the loss of fluids from the male and female reproductive systems

Integumentary system

The urinary system compensates for water loss due to sweating

The kidneys and skin both play a role in production of vitamin D

Respiratory system

The kidneys and the lungs work together to control the pH of the internal environment

Skeletal system

The kidneys and bone tissue work together to control plasma calcium levels

Digestive system

The kidneys must compensate for fluids lost by the digestive system

Muscular system

Elimination of urine from the bladder is controlled by muscle tissue

Lymphatic system

Extracellular fluid volume and composition (including lymph) are controlled by the kidneys

Nervous system

The production and elimination of urine are influenced by the nervous system

Cardiovascular system

The urinary system controls blood volume

Blood volume and blood pressure play a role in determining water and solute excretion

Endocrine system

The production and elimination of urine are influenced by the endocrine system

Urinary System
InnerConnections: Chapter 20

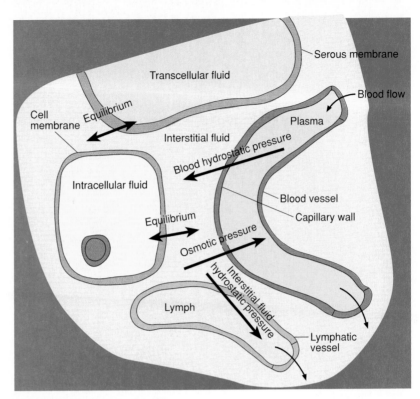

Fluid Compartments
Figure 21.4

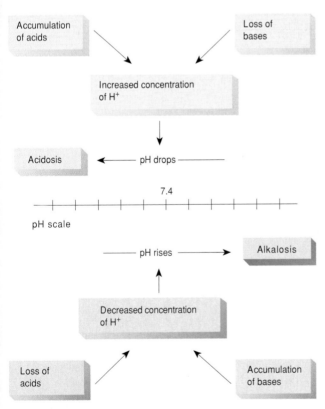

Acid-Base Imbalances
Box 21.3, Figure 21D

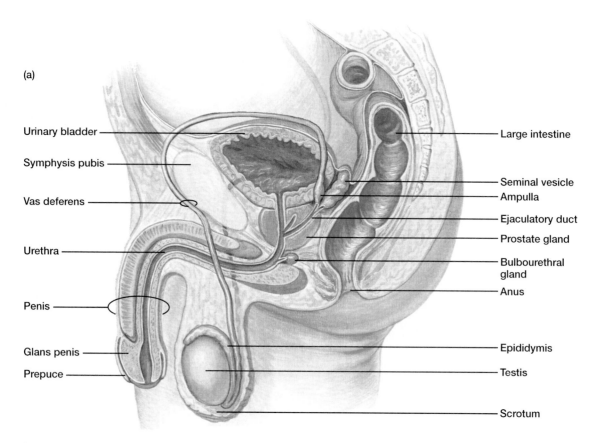

(a)

Urinary bladder

Symphysis pubis

Vas deferens

Urethra

Penis

Glans penis

Prepuce

Large intestine

Seminal vesicle

Ampulla

Ejaculatory duct

Prostate gland

Bulbourethral gland

Anus

Epididymis

Testis

Scrotum

Male Reproductive System, Sagittal
Figure 22.1a

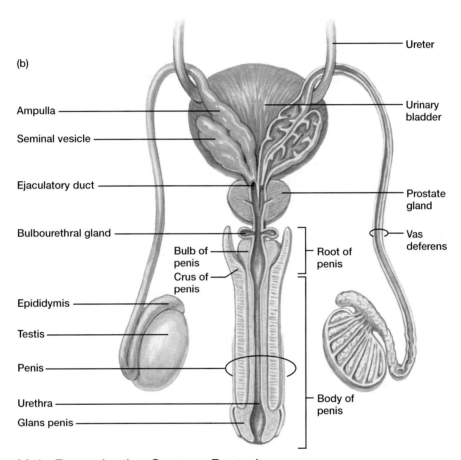

(b)

Ureter

Ampulla

Seminal vesicle

Urinary
bladder

Ejaculatory duct

Prostate
gland

Bulbourethral gland

Vas
deferens

Bulb of
penis

Root of
penis

Crus of
penis

Epididymis

Testis

Penis

Urethra

Body of
penis

Glans penis

Male Reproductive System, Posterior
Figure 22.1b

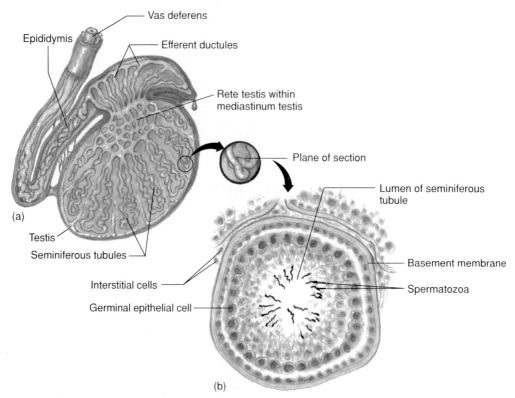

Testis and Seminiferous Tubules
Figure 22.3

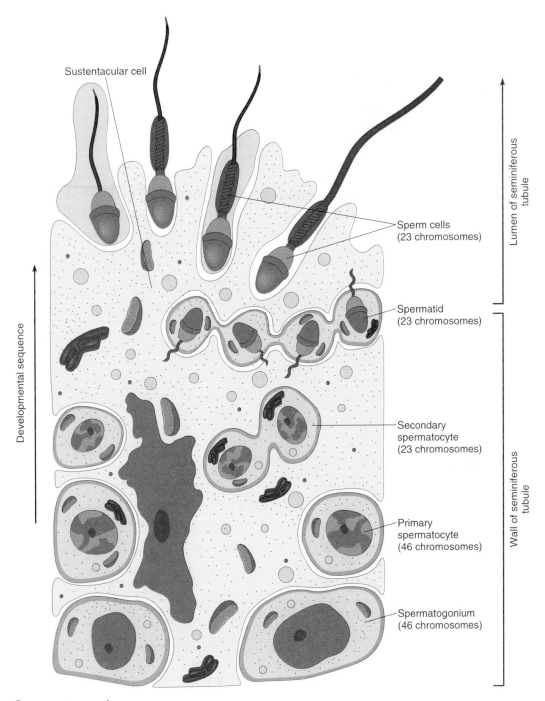

Sustentacular cell

Lumen of seminiferous tubule

Sperm cells
(23 chromosomes)

Developmental sequence

Spermatid
(23 chromosomes)

Secondary
spermatocyte
(23 chromosomes)

Wall of seminiferous tubule

Primary
spermatocyte
(46 chromosomes)

Spermatogonium
(46 chromosomes)

Spermatogonia
Figure 22.5b

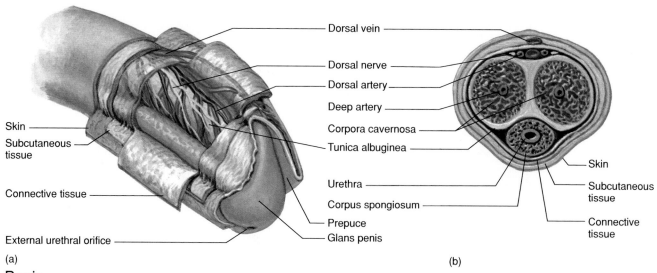

Skin

Subcutaneous tissue

Connective tissue

External urethral orifice

Dorsal vein
Dorsal nerve
Dorsal artery
Deep artery
Corpora cavernosa
Tunica albuginea
Urethra
Corpus spongiosum
Prepuce
Glans penis

Skin
Subcutaneous tissue
Connective tissue

(a)

(b)

Penis
Figure 22.11

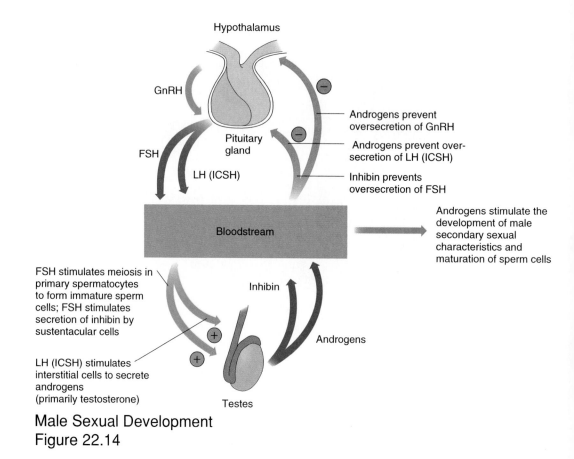

Hypothalamus

GnRH

Pituitary gland

FSH

LH (ICSH)

Androgens prevent oversecretion of GnRH

Androgens prevent over-secretion of LH (ICSH)

Inhibin prevents oversecretion of FSH

Bloodstream

Androgens stimulate the development of male secondary sexual characteristics and maturation of sperm cells

FSH stimulates meiosis in primary spermatocytes to form immature sperm cells; FSH stimulates secretion of inhibin by sustentacular cells

Inhibin

LH (ICSH) stimulates interstitial cells to secrete androgens (primarily testosterone)

Androgens

Testes

Male Sexual Development
Figure 22.14

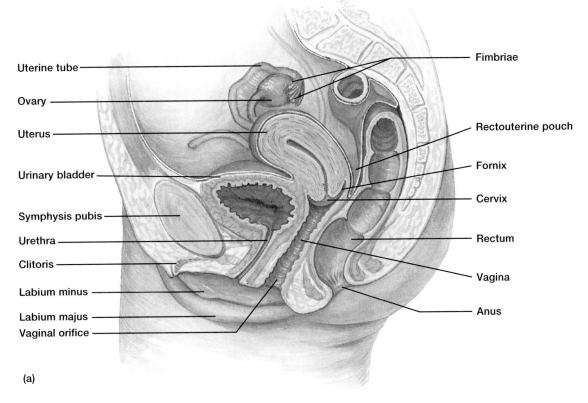

Uterine tube

Ovary

Uterus

Urinary bladder

Symphysis pubis

Urethra

Clitoris

Labium minus

Labium majus

Vaginal orifice

Fimbriae

Rectouterine pouch

Fornix

Cervix

Rectum

Vagina

Anus

(a)

Female Reproductive System
Figure 22.15a

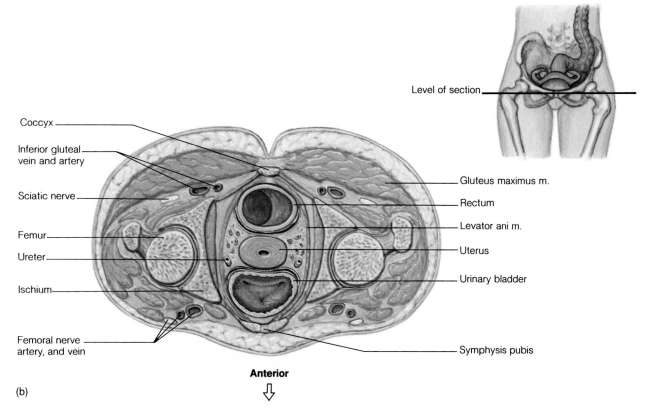

Level of section

Coccyx

Inferior gluteal
vein and artery

Sciatic nerve

Femur

Ureter

Ischium

Femoral nerve
artery, and vein

Gluteus maximus m.

Rectum

Levator ani m.

Uterus

Urinary bladder

Symphysis pubis

Anterior
⇩

(b)

Female Pelvic Cavity
Figure 22.15b

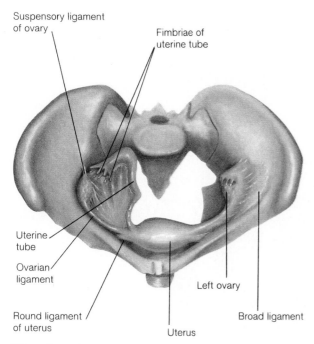

Suspensory ligament of ovary

Fimbriae of uterine tube

Uterine tube

Ovarian ligament

Round ligament of uterus

Uterus

Left ovary

Broad ligament

The Ovaries
Figure 22.16

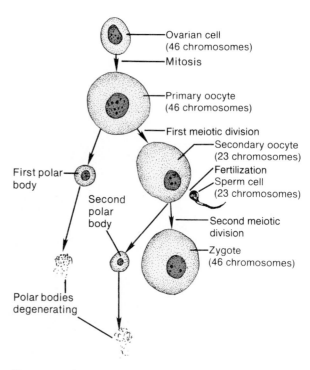

Ovarian cell (46 chromosomes)

Mitosis

Primary oocyte (46 chromosomes)

First meiotic division

Secondary oocyte (23 chromosomes)

Fertilization
Sperm cell (23 chromosomes)

Second meiotic division

Zygote (46 chromosomes)

First polar body

Second polar body

Polar bodies degenerating

Oogenesis
Figure 22.17a

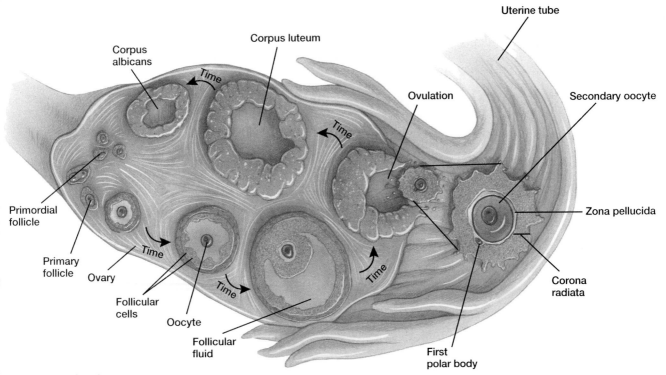

Ovarian Cycle
Figure 22.22

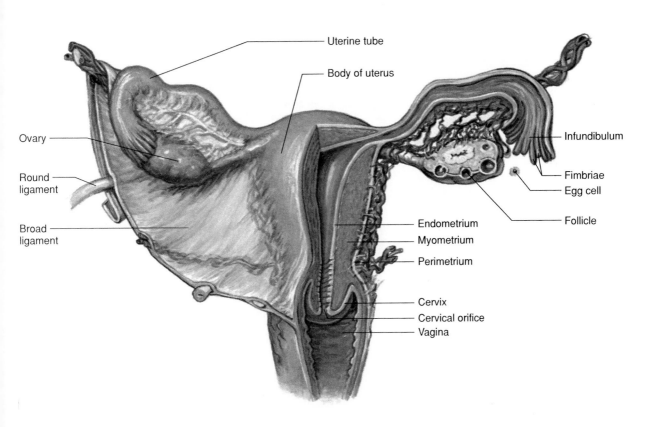

Infundibulum
Figure 22.23

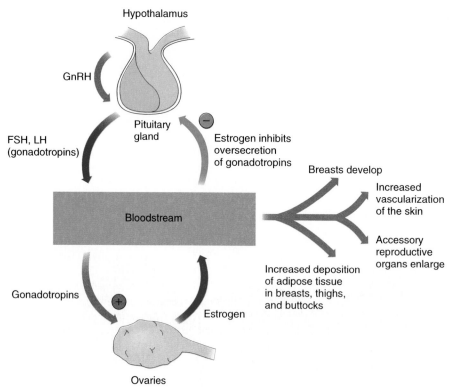

Hypothalamus

GnRH

Pituitary gland

FSH, LH (gonadotropins)

− Estrogen inhibits oversecretion of gonadotropins

Bloodstream

Breasts develop

Increased vascularization of the skin

Accessory reproductive organs enlarge

Increased deposition of adipose tissue in breasts, thighs, and buttocks

Gonadotropins

+

Estrogen

Ovaries

Female Sexual Development
Figure 22.28

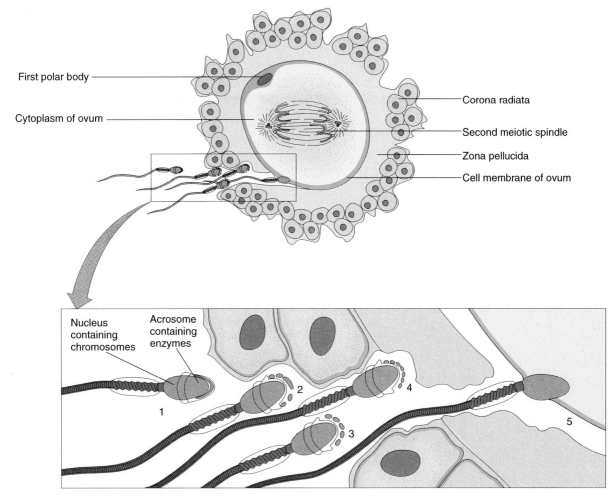

First polar body

Cytoplasm of ovum

Corona radiata

Second meiotic spindle

Zona pellucida

Cell membrane of ovum

Nucleus containing chromosomes

Acrosome containing enzymes

1

2

3

4

5

Fertilization
Figure 22.32

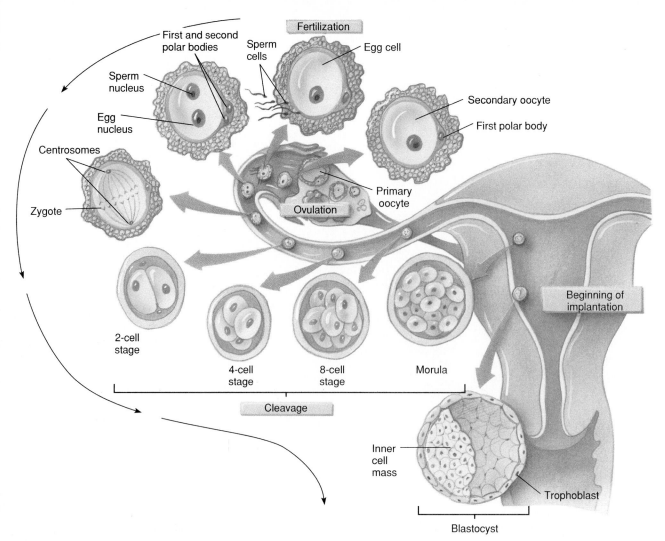

Early Human Development
Figure 22.33

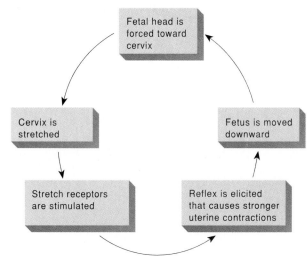

Positive Feedback Mechanism
Figure 22.36

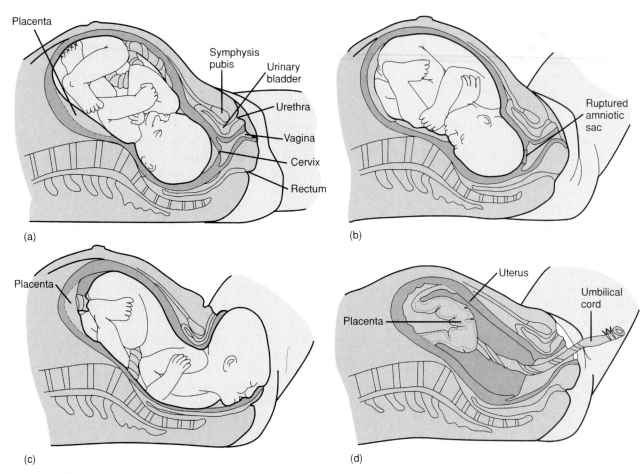

Placenta

Symphysis
pubis

Urinary
bladder

Urethra

Vagina

Cervix

Rectum

(a)

Ruptured
amniotic
sac

(b)

Placenta

(c)

Uterus

Placenta

Umbilical
cord

(d)

Stages in Birth
Figure 22.37

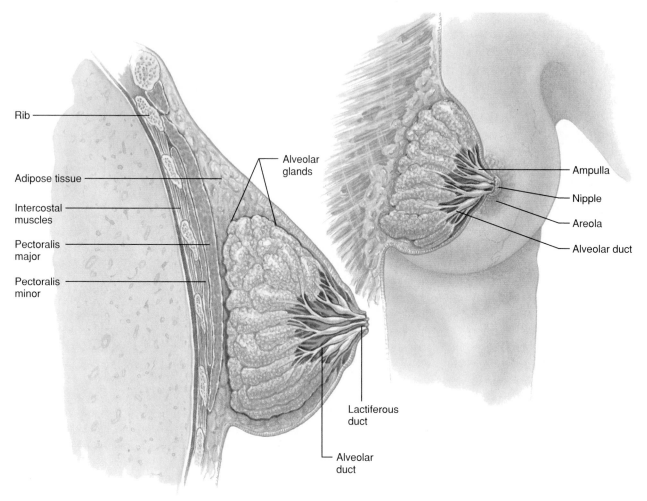

Rib

Adipose tissue

Intercostal
muscles

Pectoralis
major

Pectoralis
minor

Alveolar
glands

Lactiferous
duct

Alveolar
duct

Ampulla

Nipple

Areola

Alveolar duct

Structure of the Breast
Figure 22.38

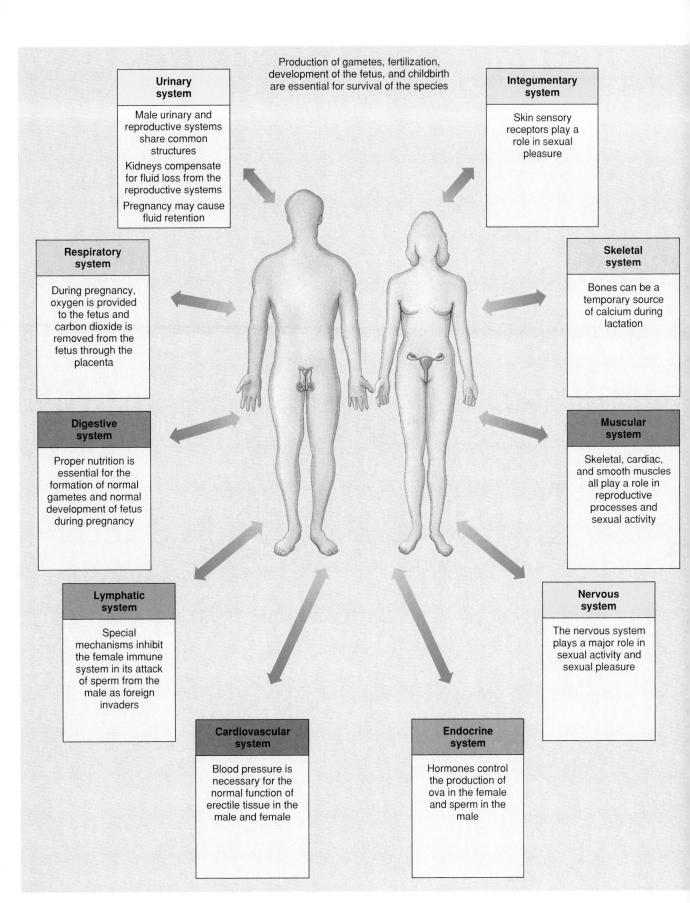

Production of gametes, fertilization, development of the fetus, and childbirth are essential for survival of the species

Urinary system

Male urinary and reproductive systems share common structures

Kidneys compensate for fluid loss from the reproductive systems

Pregnancy may cause fluid retention

Integumentary system

Skin sensory receptors play a role in sexual pleasure

Respiratory system

During pregnancy, oxygen is provided to the fetus and carbon dioxide is removed from the fetus through the placenta

Skeletal system

Bones can be a temporary source of calcium during lactation

Digestive system

Proper nutrition is essential for the formation of normal gametes and normal development of fetus during pregnancy

Muscular system

Skeletal, cardiac, and smooth muscles all play a role in reproductive processes and sexual activity

Lymphatic system

Special mechanisms inhibit the female immune system in its attack of sperm from the male as foreign invaders

Nervous system

The nervous system plays a major role in sexual activity and sexual pleasure

Cardiovascular system

Blood pressure is necessary for the normal function of erectile tissue in the male and female

Endocrine system

Hormones control the production of ova in the female and sperm in the male

Reproductive System
InnerConnections: Chapter 22

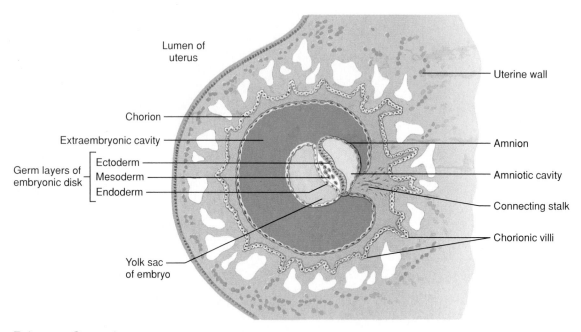

Lumen of
uterus

Chorion

Extraembryonic cavity

Germ layers of
embryonic disk

Ectoderm

Mesoderm

Endoderm

Yolk sac
of embryo

Uterine wall

Amnion

Amniotic cavity

Connecting stalk

Chorionic villi

Primary Germ Layers
Figure 23.6

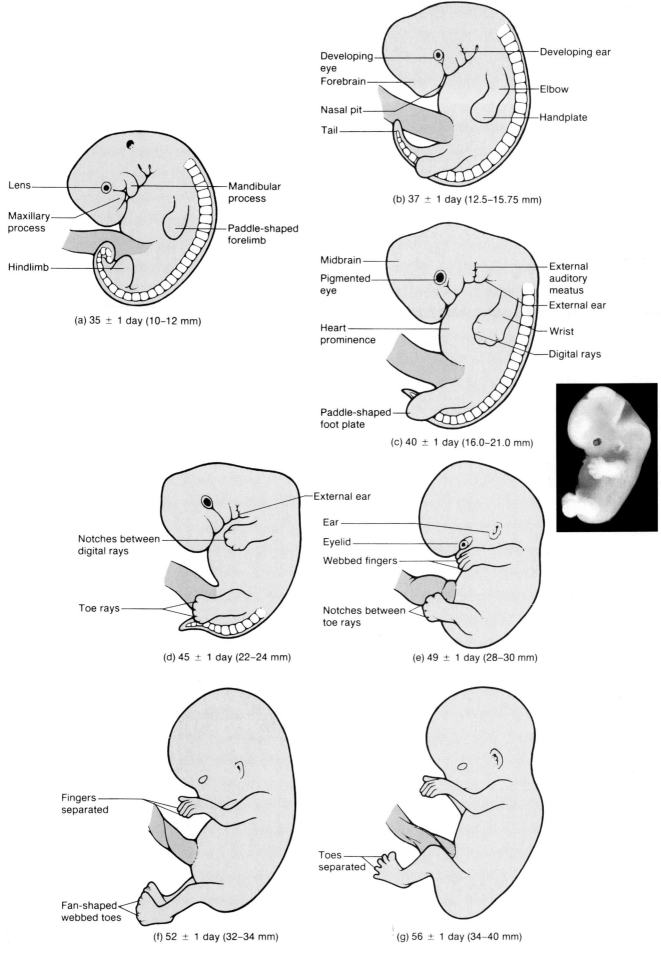

Lens

Maxillary process

Hindlimb

Mandibular process

Paddle-shaped forelimb

(a) 35 ± 1 day (10–12 mm)

Developing eye

Forebrain

Nasal pit

Tail

Developing ear

Elbow

Handplate

(b) 37 ± 1 day (12.5–15.75 mm)

Midbrain

Pigmented eye

Heart prominence

Paddle-shaped foot plate

External auditory meatus

External ear

Wrist

Digital rays

(c) 40 ± 1 day (16.0–21.0 mm)

Notches between digital rays

Toe rays

(d) 45 ± 1 day (22–24 mm)

External ear

Ear

Eyelid

Webbed fingers

Notches between toe rays

(e) 49 ± 1 day (28–30 mm)

Fingers separated

Fan-shaped webbed toes

(f) 52 ± 1 day (32–34 mm)

Toes separated

(g) 56 ± 1 day (34–40 mm)

Human Embryonic Development
Figure 23.11

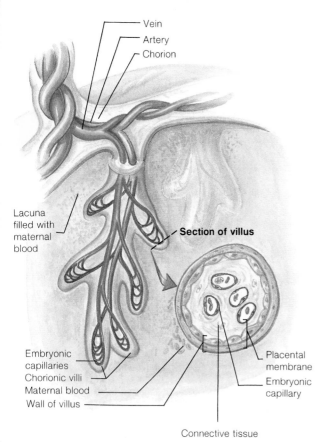

Vein
Artery
Chorion

Lacuna filled with maternal blood

Section of villus

Embryonic capillaries
Chorionic villi
Maternal blood
Wall of villus

Placental membrane
Embryonic capillary

Connective tissue

Villus
Figure 23.12

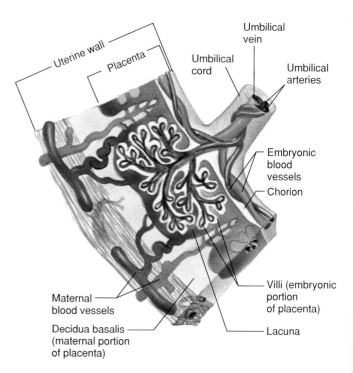

Uterine wall
Placenta

Umbilical vein
Umbilical cord
Umbilical arteries

Embryonic blood vessels
Chorion

Maternal blood vessels

Decidua basalis (maternal portion of placenta)

Villi (embryonic portion of placenta)

Lacuna

Placenta
Figure 23.13

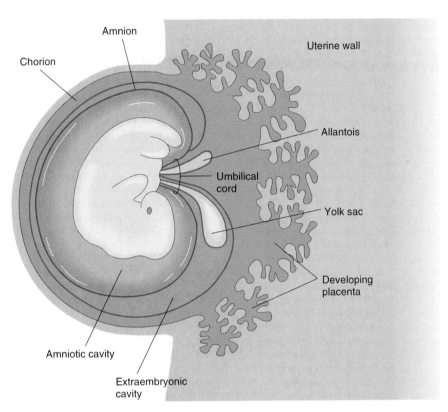

Chorion

Amnion

Uterine wall

Allantois

Umbilical
cord

Yolk sac

Developing
placenta

Amniotic cavity

Extraembryonic
cavity

Umbilical Cord
Figure 23.14

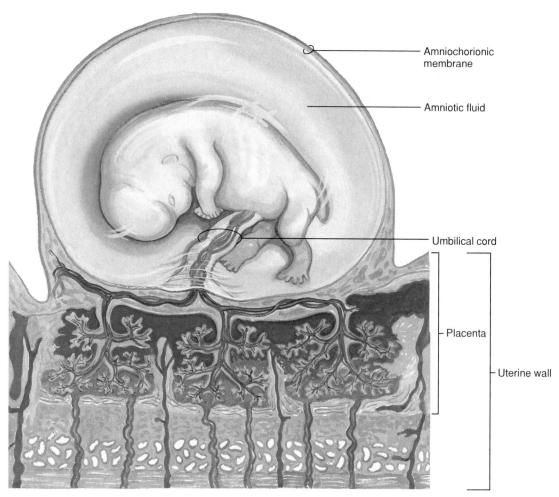

The following labels appear in the figure:

Amniochorionic membrane

Amniotic fluid

Umbilical cord

Placenta

Uterine wall

The Developing Placenta
Figure 23.15

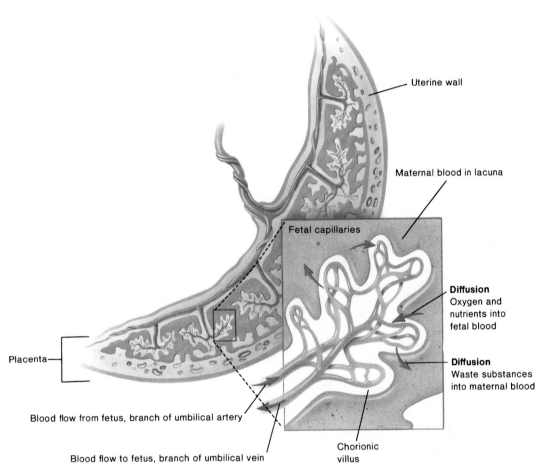

Uterine wall

Maternal blood in lacuna

Fetal capillaries

Diffusion
Oxygen and nutrients into fetal blood

Diffusion
Waste substances into maternal blood

Placenta

Blood flow from fetus, branch of umbilical artery

Blood flow to fetus, branch of umbilical vein

Chorionic villus

Maternal Blood
Figure 23.21

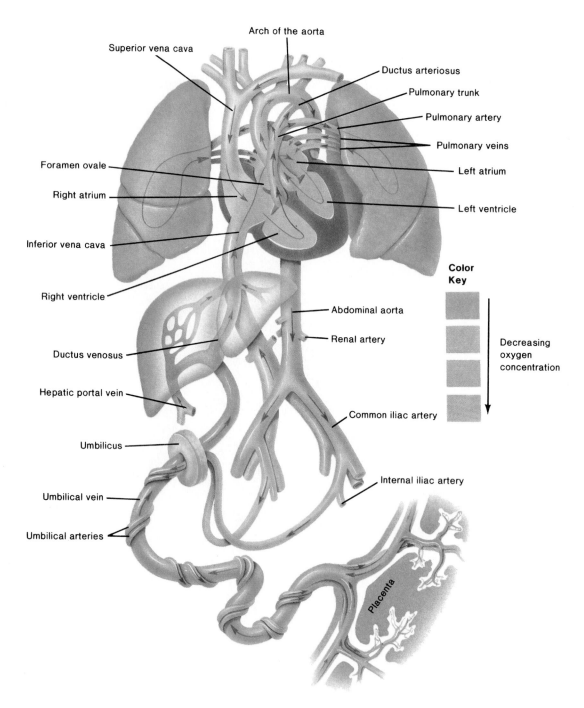

Arch of the aorta

Superior vena cava

Ductus arteriosus

Pulmonary trunk

Pulmonary artery

Pulmonary veins

Foramen ovale

Left atrium

Right atrium

Inferior vena cava

Left ventricle

Right ventricle

Abdominal aorta

Renal artery

Ductus venosus

Hepatic portal vein

Common iliac artery

Umbilicus

Internal iliac artery

Umbilical vein

Umbilical arteries

Placenta

Color Key

Decreasing oxygen concentration

Fetal Circulation I
Figure 23.22a

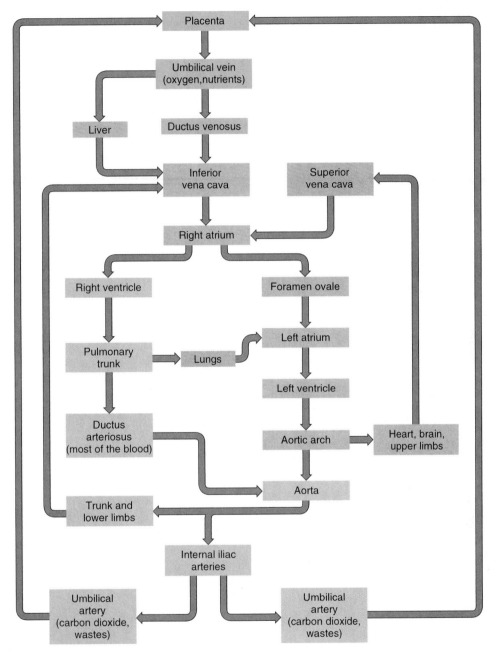

Fetal Circulation II
Figure 23.22b

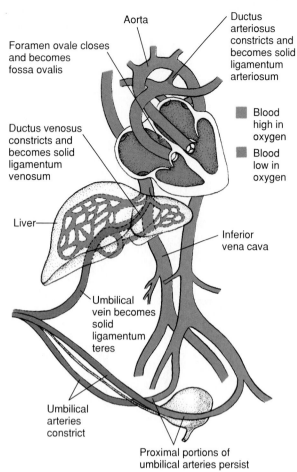

Foramen ovale closes and becomes fossa ovalis

Aorta

Ductus arteriosus constricts and becomes solid ligamentum arteriosum

Blood high in oxygen

Blood low in oxygen

Ductus venosus constricts and becomes solid ligamentum venosum

Liver

Inferior vena cava

Umbilical vein becomes solid ligamentum teres

Umbilical arteries constrict

Proximal portions of umbilical arteries persist

Newborn's Circulatory System
Figure 23.24

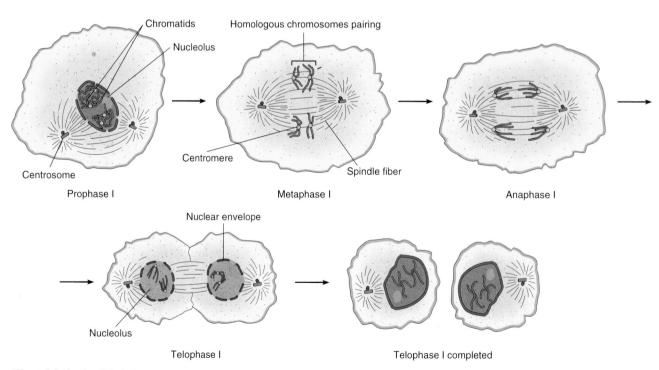

Chromatids

Nucleolus

Homologous chromosomes pairing

Centrosome

Centromere

Spindle fiber

Prophase I

Metaphase I

Anaphase I

Nuclear envelope

Nucleolus

Telophase I

Telophase I completed

First Meiotic Division
Figure 24.4

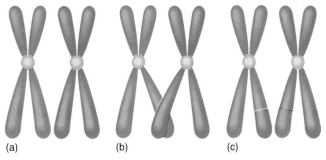

(a)   (b)   (c)

Crossing Over
Figure 24.5

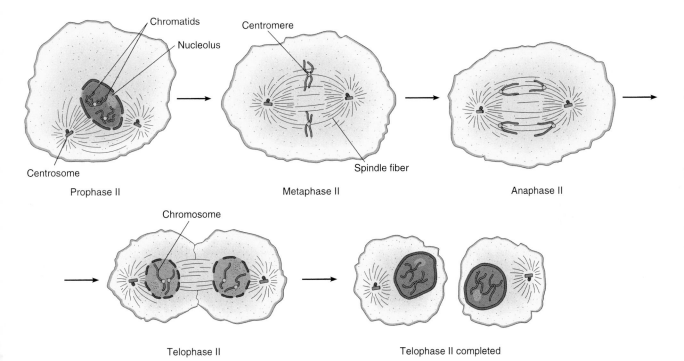

Chromatids

Nucleolus

Centrosome

Prophase II

Centromere

Spindle fiber

Metaphase II

Anaphase II

Chromosome

Telophase II

Telophase II completed

Second Meiotic Division
Figure 24.6

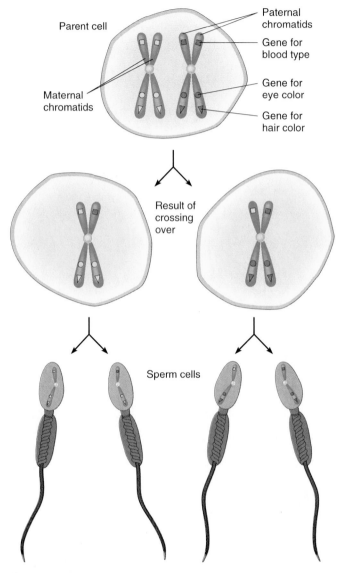

Parent cell

Paternal chromatids

Gene for blood type

Gene for eye color

Gene for hair color

Maternal chromatids

Result of crossing over

Sperm cells

Maternal and Paternal Traits
Figure 24.7

CREDITS

Photo

Figure 4.22 Courtesy of Ealing Corporation

Figure 6.3B © Ed Reschke

Figure 9.3 © Times Mirror Higher Education Group, Inc./Carol D. Jacobson, PhD., Department of Veterinary Anatomy, Iowa State University

Figure 9.11A © H. E. Huxley

Figure 10.17C © Don Fawcett/Photo Researchers, Inc.

Figure 21.11E © Dr. Landrum B. Shettles

Line Art

Line Art

Figure 3.12 From Ricki Lewis, *Life,* 2d ed. Copyright © 1995 Wm. C. Brown Communications, Inc., Dubuque, Iowa. Reprinted by permission of Times Mirror Higher Education Group, Inc., Dubuque, Iowa. All Rights Reserved.

Figure 3.28 From Stuart Ira Fox, *Human Physiology,* 4th ed. Copyright © 1993 Wm. C. Brown Communications, Inc., Dubuque, Iowa. Reprinted by permission of Times Mirror Higher Education Group, Inc., Dubuque, Iowa. All Rights Reserved.

Figure 4.16 From Ricki Lewis, *Life,* 2d ed. Copyright © 1995 Wm. C. Brown Communications, Inc., Dubuque, Iowa. Reprinted by permission of Times Mirror Higher Education Group, Inc., Dubuque, Iowa. All Rights Reserved.

Figure 5.10 From Kent M. Van De Graaff, *Human Anatomy,* 4th ed. Copyright © 1995 Wm. C. Brown Communications, Inc., Dubuque, Iowa. Reprinted by permission of Times Mirror Higher Education Group, Inc., Dubuque, Iowa. All Rights Reserved.

Figure 6.2 From Kent M. Van De Graaff and Stuart Ira Fox, *Concepts of Human Anatomy and Physiology,* 4th edition. Copyright © 1995 Wm. C. Brown Communications, Inc., Dubuque, Iowa. Reprinted by permission of Times Mirror Higher Education Group, Inc., Dubuque, Iowa. All Rights Reserved.

Figure 7.17 From Kent M. Van De Graaff, *Human Anatomy,* 3d ed. Copyright © 1992 Wm. C. Brown Communications, Inc., Dubuque, Iowa. Reprinted by permission of Times Mirror Higher Education Group, Inc., Dubuque, Iowa. All Rights Reserved.

Figure 7.19 From Kent M. Van De Graaff, *Human Anatomy,* 3d ed. Copyright © 1992 Wm. C. Brown Communications, Inc., Dubuque, Iowa. Reprinted by permission of Times Mirror Higher Education Group, Inc., Dubuque, Iowa. All Rights Reserved.

Figure 7.22 From Kent M. Van De Graaff, *Human Anatomy,* 3d ed. Copyright © 1992 Wm. C. Brown Communications, Inc., Dubuque, Iowa. Reprinted by permission of Times Mirror Higher Education Group, Inc., Dubuque, Iowa. All Rights Reserved.

Figure 7.26 From Kent M. Van De Graaff, *Human Anatomy,* 3d ed. Copyright © 1992 Wm. C. Brown Communications, Inc., Dubuque, Iowa. Reprinted by permission of Times Mirror Higher Education Group, Inc., Dubuque, Iowa. All Rights Reserved.

Figure 7.29 From Kent M. Van De Graaff, *Human Anatomy,* 3d ed. Copyright © 1992 Wm. C. Brown Communications, Inc., Dubuque, Iowa. Reprinted by permission of Times Mirror Higher Education Group, Inc., Dubuque, Iowa. All Rights Reserved.

Figure 9.20 From Kent M. Van De Graaff and Stuart Ira Fox, *Concepts of Human Anatomy and Physiology,* 4th ed. Copyright © 1995 Wm. C. Brown Communications, Inc., Dubuque, Iowa. Reprinted by permission of Times Mirror Higher Education Group, Inc., Dubuque, Iowa. All Rights Reserved.

Figure 9.21 From Kent M. Van De Graaff and Stuart Ira Fox, *Concepts of Human Anatomy and Physiology,* 4th ed. Copyright © 1995 Wm. C. Brown Communications, Inc., Dubuque, Iowa. Reprinted by permission of Times Mirror Higher Education Group, Inc., Dubuque, Iowa. All Rights Reserved.

Figure 10.4 From Stuart Ira Fox, *Human Physiology,* 3d ed. Copyright © 1990 Wm. C. Brown Communications, Inc., Dubuque, Iowa. Reprinted by permission of Times Mirror Higher Education Group, Inc., Dubuque, Iowa. All Rights Reserved.

Figure 10.6 From Kent M. Van De Graaff, *Human Anatomy,* 4th ed. Copyright © 1995 Wm. C. Brown Communications, Inc., Dubuque, Iowa. Reprinted by permission of Times Mirror Higher Education Group, Inc., Dubuque, Iowa. All Rights Reserved.

Figure 11.11 From Kent M. Van De Graaff, *Human Anatomy,* 4th ed. Copyright © 1995 Wm. C. Brown Communications, Inc., Dubuque, Iowa. Reprinted by permission of Times Mirror Higher Education Group, Inc., Dubuque, Iowa. All Rights Reserved.

Figure 11.34 From Kent M. Van De Graaff, *Human Anatomy,* 3d ed. Copyright © 1992 Wm. C. Brown Communications, Inc., Dubuque, Iowa. Reprinted by permission of Times Mirror Higher Education Group, Inc., Dubuque, Iowa. All Rights Reserved.

Figure 12.4 From Kent M. Van De Graaff and Stuart Ira Fox, *Concepts of Human Anatomy and Physiology,* 2d ed. Copyright © 1989 Wm. C. Brown Communications, Inc., Dubuque, Iowa. Reprinted by permission of Times Mirror Higher Education Group, Inc., Dubuque, Iowa. All Rights Reserved.

Figure 12.15 From Kent M. Van De Graaff and Stuart Ira Fox, *Concepts of Human Anatomy and Physiology,* 2d ed. Copyright © 1992 Wm. C. Brown Communications, Inc., Dubuque, Iowa. Reprinted by permission of Times Mirror Higher Education Group, Inc., Dubuque, Iowa. All Rights Reserved.

Figure 12.15 From Stuart Ira Fox, *Human Physiology,* 4th ed. Copyright © 1993 Wm. C. Brown Communications, Inc., Dubuque, Iowa. Reprinted by permission of Times Mirror Higher Education Group, Inc., Dubuque, Iowa. All Rights Reserved.

Figure 13.13 From Kent M. Van De Graaff and Stuart Ira Fox, *Concepts of Human Anatomy and Physiology,* 4th ed. Copyright © 1995 Wm. C. Brown Communications, Inc., Dubuque, Iowa. Reprinted by permission of Times Mirror Higher Education Group, Inc., Dubuque, Iowa. All Rights Reserved.

Figure 14.24 From Ricki Lewis, *Human Genetics.* Copyright © 1994 Wm. C. Brown Communications, Inc., Dubuque, Iowa. Reprinted by permission of Times Mirror Higher Education Group, Inc., Dubuque, Iowa. All Rights Reserved.

Figure 15.13 From Kent M. Van De Graaff, *Human Anatomy,* 4th ed. Copyright © 1995 Wm. C. Brown Communications, Inc., Dubuque, Iowa. Reprinted by permission of Times Mirror Higher Education Group, Inc., Dubuque, Iowa. All Rights Reserved.

Figure 15.22 From Kent M. Van De Graaff and Stuart Ira Fox, *Concepts of Human Anatomy and Physiology,* 4th ed. Copyright © 1995 Wm. C. Brown Communications, Inc., Dubuque, Iowa. Reprinted by permission of Times Mirror Higher Education Group, Inc., Dubuque, Iowa. All Rights Reserved.

Figure 15D From Kent M. Van De Graaff and Stuart Ira Fox, *Concepts of Human Anatomy and Physiology,* 4th ed. Copyright © 1995 Wm. C. Brown Communications, Inc., Dubuque, Iowa. Reprinted by permission of Times Mirror Higher Education Grou Inc., Dubuque, Iowa. All Rights Reserved.

Figure 15.26 From Kent M. Van De Graaff and Stuart Ira Fox *Concepts of Human Anatomy and Physiology,* 4th ed. Copyrig